**INTRODUCTORY BIOSTATISTICS
FOR THE HEALTH SCIENCES**

INTRODUCTORY BIOSTATISTICS FOR THE HEALTH SCIENCES

SECOND EDITION

Robert C. Duncan, Ph.D.
Division of Biostatistics
University of Miami School of Medicine
Miami, Florida

Rebecca G. Knapp, M.S.
Medical University of South Carolina
Charleston, South Carolina

M. Clinton Miller III, Ph.D.
Medical University of South Carolina
Charleston, South Carolina

A WILEY MEDICAL PUBLICATION
JOHN WILEY & SONS
New York · Chichester · Brisbane · Toronto · Singapore

Library of Congress Cataloging in Publication Data:

Duncan, Robert C.
  Introductory biostatistics for the health sciences.

  (A Wiley medical publication)
  Includes index.
  1. Medical statistics. 2. Biometry.   I. Knapp,
Rebecca Grant. II. Miller, M. Clinton.   III. Title.
IV. Series. [DNLM: 1. Biometry. 2. Statistics. WA 950
D913i].
RA409.D86   1983   610'.72   82-23822
ISBN 0-471-07869-7

Printed in the United States of America

10 9

PREFACE
TO THE SECOND EDITION

The first edition of this book has been used by a wide scope of students and professionals in many health-related disciplines. Our own experience and the many suggestions of others have led to what we believe to be important additions to the material. These changes should make the text more useful to a wider variety of readers.

A section has been added to the chapter on probability to allow the student to apply the concepts of sensitivity, specificity, and prevalence to the evaluation of a screening test. In response to its increasing popularity in the medical literature, new material has been added to several chapters to illustrate the use of the Bonferroni procedure for making pair-wise comparisons. New material has also been added to expand the approach to testing for differences between proportions.

The major change has been to add two new chapters on the concepts and applications of clinical and epidemiological studies. With this material the text should be appropriate for courses combining introductory biostatistics with introductory epidemiology.

We have been careful to adhere to the principles underlying the development of the first version of this text: The material should be presented in such a way that it will serve as a vehicle for self-study. This aspect will be enhanced by the inclusion of the answers to the problems.

We are grateful for the suggestions and support of many people; but, in particular, we would like to thank Dr. Alan Cantor for many helpful discussions and Ms. Debbie Gilliard for her valuable help in preparing the manuscript.

ROBERT C. DUNCAN
REBECCA G. KNAPP
M. CLINTON MILLER III

PREFACE
TO THE FIRST EDITION

The purpose of this book is to enable health science professionals to apply basic descriptive and inferential statistical techniques to problems in their field. The emphasis is on worked examples using common biomedical situations. Numerous exercises provide for mastery of the presented material. The required mathematical background is elementary algebra with an emphasis on slightly complicated algebraic summations.

The complete text is presently being used by the authors in a core course for medical students. With the elimination of Chapter 2, "Elementary Probability," it serves also as a text for a baccalaureate nursing program and several baccalaureate allied health science programs.

The content and organization of the material represent four years of experience by the authors in developing and implementing a self-paced, lecture-free course of study in statistics at the Medical University of South Carolina. The requirements of a text that will not necessarily be mediated by the traditional classroom lecture have led to several unique features. The big picture is kept in focus and trimmed to size by avoiding, as much as possible, generalized concepts and absolutely shunning peripheral issues. Mathematical arguments are avoided by referring the interested reader to advanced texts. Finally, the language and notation have been refined, through usage and review, so that they are appropriate for a statistically naive reader. Thus, the text can serve equally well for individualized study or as the basis for a traditional lecture series.

Each chapter is divided into five sections: Overview, Objectives, Foundations, Methodology, and Problems. We suggest that the student have the learning objectives clearly in mind and then read completely through the foundations and methodology sections. Then the student should work through the methodology section with references to foundations as needed. Finally, all problems should be worked to be sure the material has been mastered.

This material represents the culmination of an effort that has benefited greatly from the direct and indirect assistance of many graduate and medical students. We wish especially to thank Dr. Glenn L. Tindell for his careful reading of the manuscript and his many valuable suggestions. We also wish to express our appreciation to our typists, Ms. Barbara Wright and Ms. Pat Saranko, for their patience and forbearance.

ROBERT C. DUNCAN
REBECCA G. KNAPP
M. CLINTON MILLER III

Charleston, South Carolina
July 1976

CONTENTS

INTRODUCTORY BIOSTATISTICS FOR THE HEALTH SCIENCES

CHAPTER 1

DESCRIPTIVE STATISTICS

OVERVIEW

Statistical methodology can be separated into two main components: *descriptive* statistics and *inferential* statistics. Descriptive statistics includes the presentation of data in graphs and tables and the calculation of numerical summaries such as frequencies, averages, medians, percentages, and ranges. Inferential statistics provides a methodology for arriving at conclusions or making decisions about a population by reasoning from the evidence of observed numerical data from a sample of the population.

Generally, any set of observed data is a part of a larger aggregate of potential, but unobserved, data. The observed data are called a *sample*, whereas the larger group is called a *population*. Examples are the blood glucose levels of a group of patients (the sample) selected at random from a list of hospital outpatients known to be diabetics (the population), and a series of determinations for a set of identically prepared chemical standards (the sample) carried out in such a fashion as to be representative of all potential determinations of the particular chemical and laboratory being studied (the population).

The techniques of descriptive statistics, in the form of tabular and graphical summaries, are indispensable for organizing and understanding collections of sample data. The numerical quantities computed from sample data both describe the sample itself and provide for inferences about the characteristics of the population from which the sample was collected.

Chapter 1 develops statistical notation and presents the methods of descriptive statistics. Chapters 2 and 3 develop the concepts of probability and statistical inference. The remainder of the book is devoted to common inferential statistical methodology applied to the health sciences.

OBJECTIVES

- To construct an absolute frequency table and a relative frequency table from a set of raw data
- To present a frequency table graphically by means of a histogram
- To perform simple manipulations using $\sum$ notation
- To compute measures of central tendency—mean, median, and mode
- To compute measures of variability—range, variance, and standard deviation
- To choose the appropriate descriptive statistics for summarizing a set of data

FOUNDATIONS

Levels of Measurement

All data may be classified according to four measurement levels or measurement "scales." Since the type of measurement scale influences the choice of statistical analysis, we must consider the different forms in which medical data can occur. First, we will define the four measurement scales, then introduce the techniques for describing statistical data, and finally discuss how the measurement scale dictates which descriptive technique is appropriate.

Measurement at its simplest level exists when a "name," a number, or other symbol is used to assign subjects to specified categories of a given variable. For example, measurement of the variable "blood type" consists of the classifications "type O," "type A," "type B," and "type AB." Similarly, for the variable "psychiatric diagnosis," the measurement assigned to each patient may consist of one of the following numbers corresponding to specified diagnostic groups:

1. Paranoid
2. Schizophrenic
3. Manic-depressive
4. Psychoneurotic

When names, numbers, or other symbols are used, as in the above examples, to specify the groups to which various subjects belong, the measurement scale is called the *nominal* or classificatory scale. Other examples of variables measured on the nominal scale are sex, race, and marital status. The subclasses composing the categories of the nominal scale must be exhaustive and mutually exclusive; every individual must fall into one and only one category. For the nominal scale, there is no necessary relationship between the categories or subclasses of the variable being measured. Within any one subclass, the members are assumed to be equivalent with respect to the characteristic being scaled. In addition, the names or symbols designating the subclasses may be interchanged without altering the essential information conveyed by the scale.

The second type of measurement scale, the *ordinal scale*, differs from the nominal scale in that the different categories or subclasses specified in the scale are ranked in terms of a graded order (greater than, less than, equal to). Classification of patients according to the variable "patient satisfaction" into the subclasses "very satisfied," "moderately satisfied," "moderately unsatisfied," and "very unsatisfied" constitutes an ordinal scale of measurement. Other examples include patient condition (1 = "unimproved," 2 = "stable," 3 = "improved") and class standing (upper third, middle third, lower third). It does not matter what symbols are assigned to represent the subclasses so long as the ordering or ranking of the classes is preserved. For example, without any loss of information, the numbers representing patient condition could be -2 = "unimproved," 0 = "stable," and $+2$ = "improved." Note, however, that while it is known that a patient classified as "unimproved" is more ill than a patient classified as "stable," we do not know how much more ill he is. Also, the difference in illness status from "unimproved" to "stable" is not necessarily the same as the difference in status from "stable" to "improved." Note that the names assigned to the categories of an ordinal scale have to be in the proper numerical order (i.e., either from low to high or from high to low) so that the subjects are *ranked* according to their scores.

When the distances between any two numbers on a scale are known units of equal size, then the level of measurement achieved by the data is the *interval scale*. With this scale it is possible not only to declare that one subclass represents "more" or "less" of the variable being measured than another subclass, but it may also be determined exactly "how much" more. Temperature in degrees Fahrenheit or Celsius is an example of a variable measured on an interval scale. The difference in temperature between 100°F and 104°F is the same as the difference in temperature between 50°F and 54°F. One notable shortcoming of the interval scale, however, is the absence of a true zero point; that is, the zero point on the scale does not represent the true or theoretical absence of the quantity being measured. For example, 0°C is simply the point at which water freezes. It does not represent the true absence of temperature. When measurement begins at a true zero point and the scale also has equal intervals, then the *ratio scale* of measurement has been achieved. Examples of variables measured on the ratio scale are length, time, mass, volume, and temperature in degrees Kelvin. The interval scale, while allowing the ordinary operations of arithmetic, does not permit the formation of ratios. Since there is no true zero point on the scale, it may not be said that 4°C represents four times as much heat as 0°C. In contrast, in a true ratio scale, such as one using degrees Kelvin, 50°K represents exactly twice as much temperature as 25°K.

The four types of measurement scales—the nominal, the ordinal, the interval, and the ratio scales—represent increasing orders of sophistication in the measurement process. As stated earlier, the type of analysis appropriate in a given situation depends in part upon the level of measurement achieved by the data. Data that are measured on the nominal or ordinal scale require, in general, *nonparametric* statistical analyses, a relatively new branch of inferential statistics. A statistical technique appropriate for nominal or ordinal data will be presented in Chapter 7. The classical parametric statistical techniques, which require in general at least an interval scale of measurement, are discussed in Chapters 3–7. The choice of appropriate descriptive techniques is also influenced by level of measurement of the data. We will return to a discussion of the topic in a later section of this chapter.

Frequency Tables and Frequency Diagrams

The first step in the analysis and interpretation of a set of sample data is the reduction of the observed numbers to a set of descriptive statistics that display important features of the sample. The sample can be described either pictorially in terms of graphs and charts or quantitatively in tables of numerical summaries.

The simplest, and often most useful, summary of a sample is a table of the frequency with which individual values were found in the sample. The list of frequencies also can be presented in graphical form as a frequency diagram.

Table 1.1.   Scores of 12 psychiatric patients on a 5-point anxiety scale

Patient:	1	2	3	4	5	6	7	8	9	10	11	12
Anxiety score:	4	3	5	1	4	4	2	5	4	3	4	5

These concepts will be illustrated with the data in Table 1.1, which represents the measurement of level of anxiety on a sample of 12 psychiatric patients. By tabulating the number of times each score was obtained, or the *frequency* of each score, as in Table 1.2, various features of the sample are made evident. Every possible score is represented at least once and the most frequently obtained score was 4; the scores are skewed toward the higher end of the scale.

An alternate presentation that conveys the same information is the frequency diagram shown in Figure 1.1. The frequency diagram imparts a sense of the "shape" of the distribution of frequencies among the possible

Table 1.2. Frequency distribution of anxiety scores from the data in Table 1.1

Score	Frequency
1	1
2	1
3	2
4	5
5	3
Total	12

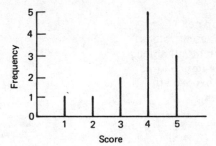

Figure 1.1. Frequency diagram of anxiety scores from the data in Table 1.1.

scores. An important feature of the shape of a frequency diagram, or the frequency distribution, is that of symmetry or lack of symmetry. A *symmetric* distribution is one in which frequencies are equal (or nearly equal) at points on either side of a central value. If a frequency distribution is not symmetric, it is said to be *skewed.* Stylized examples of symmetric and skewed distributions are shown in Figure 1.2.

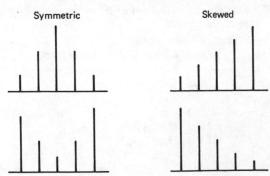

Figure 1.2. Frequency diagrams of symmetric and skewed distributions.

Frequency Tables and Frequency Diagrams for Grouped Data

When the number of distinct values in a sample is small, as in Table 1.1 where the data can have only the values 1 through 5, a simple frequency table and the associated frequency diagram can be used. Whenever the number of distinct sample values is large, the data can be divided into groups so that the sample values within a group are similar to each other. Then the frequency of observations within the groups can be tabulated instead of the frequency of individual values.

The specific details for constructing grouped frequency tables are presented in the section on methodology. The concept is illustrated in Tables 1.3 and 1.4. From Table 1.4 it is immediately clear that the distribution of sample values is symmetric about the interval 9.95–11.95 μ mole/min/ml

Table 1.3. Red blood cell cholinesterase values (μ mole/min/ml) among thirty-five agricultural workers exposed to pesticides

10.6	12.5	11.0	9.2	11.6
9.9	11.8	11.6	15.3	12.6
12.6	12.4	12.2	10.9	16.7
15.2	10.2	13.4	9.0	7.7
12.3	11.3	9.9	11.0	10.9
11.7	9.4	9.8	8.6	10.1
12.3	11.4	10.2	12.5	8.7

Table 1.4. Grouped frequency distribution for the data in Table 1.3

RBC Cholinesterase (μ mole/min/ml)	Frequency
5.95–7.95	1
7.95–9.95	8
9.95–11.95	14
11.95–13.95	9
13.95–15.95	2
15.95–17.95	1
Total	35

and also that this is the most frequent range of values. The sample mean and the median also should lie in this interval.

The frequency diagram for grouped data is called a *histogram*. The histogram for the RBC cholinesterase data, Figure 1.3, emphasizes graphically the features observed in Table 1.4.

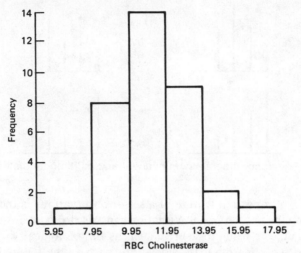

Figure 1.3.   Histogram of the frequency distribution of the data in Table 1.3.

Numerical Summaries of Sample Data

Quantitative descriptions of samples usually involve measures of location, or central tendency, and measures of spread, or dispersion. The three common measures of central tendency are the mean, the median, and the mode. The amount of spread in a set of data usually is measured by either the range of observations or the standard deviation. Details of the calculation of these quantities will be found in the section on methodology.

Location.   The *mean* of a sample is the arithmetic average of the sample values. It represents the center of the data according to the size of the values. The mean is sometimes called the "center of gravity" or the "expected value."

The *median* divides the sample so that half the observations are below it and half of them are above it. Thus, the median locates the center of the data by count and disregards size.

The *mode* is related to the concept of a peak or peaks in the frequency distribution. If there is only one peak, the distribution is said to be unimodal and the sample value corresponding to this peak is called the mode. This corresponds to the definition of the mode as the most frequent sample value. In

Figure 1.1 the sample score of 4 is seen to be the mode of the data in Table 1.1. When there are two or more peaks in the frequency diagram, the distribution is said to be multimodal. Various frequency diagrams are illustrated in Figure 1.4.

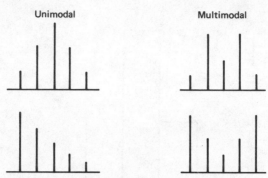

Figure 1.4. Frequency diagrams of unimodal and multimodal distributions.

Variation. An important feature of a set of observations is the amount of spread among the sample values. Variation can arise from intrinsic differences among the subjects being measured, such as the biological variation from patient to patient, or from extrinsic factors, such as the failure of the measuring device to yield the same result when measuring the same subject over and over again. Together these two sources of variation combine to determine the spread among observed values.

Two common measures of variation are the range of sample values and the standard deviation of the sample. The computation of these quantities and choice of an appropriate measure will be taken up in the following section after suitable notation has been developed.

METHODOLOGY

Construction of Frequency Tables for Grouped Data

Interval construction. The first step in constructing a frequency table is to order the sample observations by size. Table 1.5 shows the data in Table 1.3 ordered from smallest to largest value.

The next step is to determine the number and location of the intervals to be used in the frequency table. Usually, 6 to 12 intervals will suffice. Too few intervals will obscure details and too many intervals will defeat the purpose of grouping. The range in Table 1.5 is 9 units (16.7 – 7.7) so the data could

Table 1.5. Ordered observations for the data in Table 1.3

7.7	9.9	10.9	11.7	12.5
8.6	9.9	11.0	11.8	12.6
8.7	10.1	11.0	12.2	12.6
9.0	10.2	11.3	12.3	13.4
9.2	10.2	11.4	12.3	15.2
9.4	10.6	11.6	12.4	15.3
9.8	10.9	11.6	12.5	16.7

be contained in 9 intervals of width 1, 5 to 6 intervals of width 2, and so on. The location of the intervals is also somewhat arbitrary. The first interval could be from 6 to 8 or from 7 to 9. It may be necessary to try several combinations until the preferred presentation is found.

Once interval width and location are chosen, the end-points, called the true limits, of the intervals must be specified in such a way that every observation can be assigned to one and only one interval. To do this the boundaries of an interval are specified to be 5 in the decimal place following the last significant digit in the data.

The first interval in Table 1.5 was to include all observations equal to or greater than 6.0 but less than 8.0 (the sample data were recorded to one decimal place) so the true interval limits are 5.95 to 7.95. If a sample of ages was recorded as "age in years to the nearest birthday" and a frequency table with decade intervals constructed, the true limits of one such interval might be 19.5–29.5. Financial data reported to the nearest thousand dollars might have an interval such as $1500–$4500 to contain observations greater than $1000 but less than $5000.

Relative and cumulative frequency. Frequency tables are usually easier to interpret and to compare if the number of observations falling into each interval is presented as a fraction or percentage of the total number of observations. The *relative* frequency is, by definition, the absolute frequency divided by the total number of observations. The relative frequency in percent is obtained by multiplying the relative frequency by 100.

The *cumulative* frequency is the fraction or percentage of observations below the upper boundary of each successive interval. It is obtained by adding the relative frequencies.

Table 1.6 shows the data in Table 1.4 with relative and cumulative frequency columns added. Note that the relative frequency and percent relative frequency columns add up to approximately 1.0 and 100.0, respectively. This provides a check on the computations. From Table 1.6 we see that 40 percent of the observations were equal to or greater than 10.0 and less

Table 1.6. Grouped frequency distribution for the data in Table 1.4, showing relative and cumulative frequency columns

RBC Cholinesterase (mole/min/ml)	Frequency	Relative Frequency	Relative Frequency (%)	Cumulative Frequency (%)
5.95–7.95	1	.029	2.9	2.9
7.95–9.95	8	.229	22.9	25.8
9.95–11.95	14	.400	40.0	65.8
11.95–13.95	9	.257	25.7	91.5
13.95–15.95	2	.057	5.7	97.2
15.95–17.95	1	.029	2.9	100.1
Total	35	1.001	100.1	

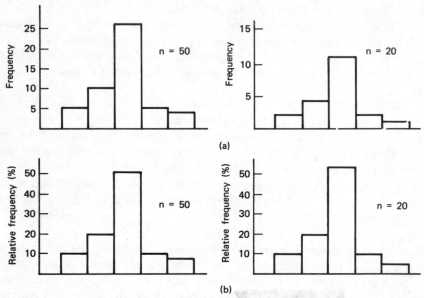

Figure 1.5. (a) With each histogram plotted on the basis of its own absolute frequencies the vertical scales are not comparable. The visual impression is misleading and it is difficult to make comparisons. (b) When plotted against relative frequency the histograms appear to be almost identical, reflecting little difference between the samples in terms of percentage distribution.

than 12.0 μ mole/min/ml. Also, 91.5 percent of the observations were less than 14.0 μ mole/min/ml.

Note that grouping data into intervals causes the identity (i.e., value) of individual data points to be lost. In a sense we are paying for clarity of presentation with the loss of "information" relative to individual values. The maximum information is contained in the raw data. Any statistical procedure reduces this information but highlights features of the data that can be readily used.

Histograms using relative frequency. When the intervals of a frequency table are of uniform width, the corresponding histogram can be plotted with relative frequency as the vertical axis. This is especially helpful when comparing two frequency distributions based on different numbers of total observations. Figure 1.5 illustrates this point.

Numerical Descriptive Measures: Methods of Computation

Sigma notation. Before beginning the discussion of descriptive measures, we must first review the use of the mathematical symbol $\sum$ (sigma), one of the most frequently used symbols in statistics. The presence of $\sum$ in a formula indicates that a group of numbers is to be added or summed. In general, the summation operation may be expressed

$$\sum_{i=1}^{n} Y_i = Y_1 + Y_2 + \cdots + Y_n \tag{1.1}$$

where $\sum$ is the summation operator, Y_i is the variable to be summed, i is the index of summation, and the 1 and n appearing below and above the symbol $\sum$ designate the range of summation. Thus, $\sum_{i=1}^{n}$ says "add all the observed values of a variable whose subscripts are between 1 and n, inclusive." Given the set of numbers $Y_1 = 1$, $Y_2 = 3$, $Y_3 = 2$, $Y_4 = 4$, $Y_5 = 5$, study the following examples carefully:

$$\sum_{i=1}^{5} Y_i = Y_1 + Y_2 + Y_3 + Y_4 + Y_5 = 1 + 3 + 2 + 4 + 5 = 15 \tag{1.2}$$

$$\sum_{i=2}^{4} Y_i = Y_2 + Y_3 + Y_4 = 3 + 2 + 4 = 9 \tag{1.3}$$

$$\sum_{i=1}^{5} Y_i^2 = Y_1^2 + Y_2^2 + Y_3^2 + Y_4^2 + Y_5^2$$
$$= (1)^2 + (3)^2 + (2)^2 + (4)^2 + (5)^2 = 55 \tag{1.4}$$

$$\left(\sum_{i=1}^{5} Y_i\right)^2 = (Y_1 + Y_2 + Y_3 + Y_4 + Y_5)^2$$
$$= (1 + 3 + 2 + 4 + 5)^2 = (15)^2 = 225 \tag{1.5}$$

Note: $\sum Y_i^2 \neq (\sum Y_i)^2$

$$\sum_{i=1}^{5} (Y_i - 3) = (1 - 3) + (3 - 3) + (2 - 3)$$
$$+ (4 - 3) + (5 - 3)$$
$$= (-2) + (0) + (-1) + (1) + (2) = 0 \tag{1.6}$$

$$\sum_{i=1}^{5} (Y_i - 3)^2 = (1 - 3)^2 + (3 - 3)^2 + (2 - 3)^2$$
$$+ (4 - 3)^2 + (5 - 3)^2$$
$$= (-2)^2 + (0)^2 + (-1)^2 + (1)^2 + (2)^2 = 10 \tag{1.7}$$

$$\sum_{i=1}^{5} 3Y_i = 3(1) + 3(3) + 3(2) + 3(4) + 3(5)$$
$$= 3(1 + 3 + 2 + 4 + 5) = 3 \sum Y_i \tag{1.8}$$

$$\sum_{i=1}^{n} (2Y_i + 2X_i) = 2 \sum_{i=1}^{n} Y_i + 2 \sum_{i=1}^{n} X_i \tag{1.9}$$

In the special case where all of the Y_i are equal, say $Y_i = c$, we have

$$\sum_{i=1}^{n} Y_i = Y_1 + Y_2 + \cdots + Y_n$$
$$= c + c + \cdots + c$$
$$= nc$$

Or, as it is commonly written,

$$\boxed{\sum_{i=1}^{n} c = nc} \tag{1.10}$$

Thus, equation (1.6) may be written as

$$\sum_{i=1}^{5} (Y_i - 3) = \sum_{i=1}^{5} (Y_i) - \sum_{i=1}^{5} 3$$
$$= 15 - 5(3)$$
$$= 0. \tag{1.6'}$$

You should understand these examples and be able to make up your own.

Mean. In sigma notation the mean of a sample is defined as

$$\bar{Y} = \frac{1}{n} \sum_{i=1}^{n} Y_i \tag{1.11}$$

where n is the number of observations in the sample and the Y_i are the sample values.

For the example in the preceding section

$$\bar{Y} = \frac{1}{5} \sum_{i=1}^{5} Y_i$$

$$= \frac{1}{5}(1 + 3 + 2 + 4 + 5)$$

$$= \frac{1}{5}(15)$$

$$= 3$$

An important property is that if the mean is subtracted from all sample values, the sum of these differences is zero.

The difference between a sample value and the sample mean is called a deviation. The ith deviation is defined as $Y_i - \bar{Y}$. Thus,

$$\sum_{i=1}^{n} (Y_i - \bar{Y}) = 0 \tag{1.12}$$

The method of proof and the verification for the sample in the preceding section is shown by equation (1.6′).

When n is small the mean is very sensitive to extreme values. If the observation whose value is 5 in the sample 1, 3, 2, 4, 5 is replaced by 50, the mean changes from 3 to 12. If it is replaced by 500, the mean changes to 102, and so on. The mean reflects the size of sample values.

Median. Consider the set of observations (15, 1, 3, 4, 1, 12, 8, 13, 9). To find the median, the observations are first arranged in order of magnitude (1, 1, 3, 4, 8, 9, 12, 13, 15). For an odd n, the median is the middle number, or here median = 8. Note that if the observation equal to 15 were replaced by 150 or even 15 million, the median would not change. It is not sensitive to extreme values.

For an even number of observations, the median is the average of the two middle values. Consider the set (10, 1, 7, 11, 15, 6). Arranged in order of magnitude, we have (1, 6, 7, 10, 11, 15). The median would be half way between 7 and 10, or median = 8.5.

The median divides a distribution into equal parts by *count* so that 50 percent of data points are above the median and 50 percent are below.

⚹ The median is useful for summarizing skewed (not symmetric) data since it is not sensitive to extreme values. For skewed data, the mean may give a misleading picture. For symmetric distributions, the median and the mean are equal. Note that neither the mean nor the median is necessarily equal to one of the sample values.

Mode. When the frequency of one value in a frequency distribution is larger than the frequencies of the other values, or when the frequency of one group in a grouped frequency distribution is larger than that for any other group, that value or group is the mode of the distribution.

For the data (1, 4, 3, 1, 2, 5) the mode is 1. For the set (2, 4, 2, 3, 1, 5, 1) the values 1 and 2 occur with equal frequency. In this case the distribution is *bimodal*, that is, has two modes.

The modal class, or group, for the data in Table 1.6 is the interval 9.95–11.95 μ mole/min/ml.

Measures of variability. Figure 1.6 represents histograms of two distributions with equal mean values but a different spread of sample values around the mean.

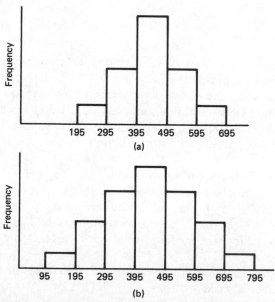

Figure 1.6. Both distributions appear to have the same mean value while (b) shows a wider range of dispersion than does (a).

Range. The *range* is the difference between the largest and smallest values in the sample. In Figure 1.6a the range would be approximately $695 - 195 = 500$, whereas in b it would be approximately $795 - 95 = 700$.

Because of its ease of computation, the range is useful as a "rough and ready" measure of variability. However, since only two numbers in the data set are utilized in its calculation, the range may give a misleading impression of the true variability of the data. This would be especially true if most of the sample were closely clumped except for one or two extreme observations.

To see this, refer to Table 1.5. If the sample value of 7.7 were changed to 2.7, the range would change from 9 to 14 but the mean and the shape of the distribution would change very little. A more stable measure of the spread among sample values is the standard deviation.

Variance, standard deviation. The deviations of the individual sample values from the sample mean give an indication of how spread out the sample is. The larger the deviations, the more dispersed the sample values are from their center, the mean.

Since the sum of the deviations is zero, a straight averaging of deviations will not serve as a measure of spread. Instead, the deviations are first squared, then averaged, and the square root is used to represent the "standard" deviation of the sample.

The procedure is illustrated using the data in Table 1.7. The quantity $\sum(Y_i - \bar{Y})^2$ is referred to as the *sum of squares*. The *variance*, which represents the average of the squared deviations, is denoted by s^2. It is computed

Table 1.7. Weight gain for six rats with supplemental diet

Rat (i)	Wt. Gain (Y_i)	Deviations $(Y_i - \bar{Y})$	$(Y_i - \bar{Y})^2$
1	6	3	9
2	2	-1	1
3	4	1	1
4	1	-2	4
5	3	0	0
6	2	-1	1
	$\sum Y_i = 18$	$\sum(Y_i - \bar{Y}) = 0$	$\sum(Y_i - \bar{Y})^2 = 16$

$$\bar{Y} = \frac{\sum_{i=1}^{6} Y_i}{n} = \frac{18}{6} = 3.0$$

as

$$s^2 = \frac{\sum(Y_i - \bar{Y})^2}{n - 1} = \frac{16}{5} = 3.2 \tag{1.13}$$

For mathematical reasons, $n - 1$ is used instead of n in computing s^2. The standard deviation, denoted by s, is

$$s = +\sqrt{s^2} = \sqrt{3.2} = 1.79 \tag{1.14}$$

For ease of computation, the sum of squares, $\sum(Y_i - \bar{Y})^2$ may be rewritten, using equations (1.2)–(1.10), as

$$\sum(Y_i - \bar{Y})^2 = \sum(Y_i^2 - 2\bar{Y}Y_i + \bar{Y}^2)$$
$$= \sum Y_i^2 - 2\bar{Y}\sum Y_i + \sum \bar{Y}^2$$

Since $\bar{Y}$ is a constant, by equation (1.10), $\sum \bar{Y}^2 = n\bar{Y}^2$. Also, putting $\bar{Y} = \frac{1}{n}\sum Y_i$,

$$\sum(Y_i - \bar{Y})^2 = \sum Y_i^2 - 2\frac{\sum Y_i}{n}\sum Y_i + n\left(\frac{\sum Y_i}{n}\right)^2$$

$$= \sum Y_i^2 - 2\frac{(\sum Y_i)^2}{n} + \frac{(\sum Y_i)^2}{n}$$

$$= \sum Y_i^2 - \frac{(\sum Y_i)^2}{n} \tag{1.15}$$

Thus, the computational formula for standard deviation is

$$s = \sqrt{\frac{\sum Y_i^2 - \frac{(\sum Y_i)^2}{n}}{n - 1}} \tag{1.16}$$

Rewriting Table 1.7, the computations are simplified as follows:

Rat (i)	Wt. Gain Y_i	Y_i^2
1	6	36
2	2	4
3	4	16
4	1	1
5	3	9
6	2	4
	$\sum Y_i = 18$	$\sum Y_i^2 = 70$

The variance of the six observations is

$$s^2 = \frac{\sum Y_i^2 - \frac{(\sum Y_i)^2}{n}}{n-1} = \frac{70 - \frac{(18)^2}{6}}{5} = \frac{16}{5} = 3.2$$

as before.

Although we probably use the concept of variability more often in daily life than we do mean values, it is a very difficult topic to master in a formal way. If you feel uncomfortable as you study this section, do not be alarmed. Simply follow the computational rules and, through repetition, the use of such measures as the range and the standard deviation will become natural and meaningful.

Choosing Appropriate Descriptive Statistics

An important factor in the selection of an appropriate descriptive measure is the level of measurement achieved by the data. For *nominal data,* as would be expected, there are few admissible descriptive statistics. Even if numerical codes are used to define categories, it is not appropriate to perform numerical operations on nominal scale variables. The only summary that is appropriate is the frequency of observations in the various categories and the mode. For example, consider Table 1.8, which presents information on diagnoses of 100 patients who visited a mental health clinic (a nominal variable). A graphical representation of the frequency counts of nominal data may be accomplished by means of a *bar chart*. A bar chart is similar to the histogram described earlier except that the horizontal scale is a group of distinct categories rather than continuous numerical intervals. Figure 1.7 shows the frequency distribution of the data given in Table 1.8. As can be seen from Figure 1.7, the modal diagnosis is "psychosis," since this is the most frequently occurring among the four categories.

Table 1.8. Frequency of diagnosis in 100 patients visiting a mental health clinic

Diagnosis	Frequency	Relative Frequency (%)
Organic brain syndrome	20	20
Psychosis	35	35
Mental retardation	20	20
Personality disorder	25	25
	100	100

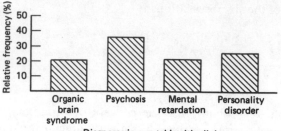

Figure 1.7. Bart chart of frequency of diagnoses in mental health clinic.

The descriptive statistics most appropriate for describing *ordinal data* are frequency counts, the median, and the mode. Suppose, for example, that 20 patients in a geriatric nursing care center are asked to rate quality of care on a scale consisting of the following categories: 1 = very poor, 2 = poor, 3 = fair, 4 = good, 5 = very good. The frequency table for these data may be as shown in Table 1.9. As for nominal data, the frequency counts for Table 1.9 may be shown graphically by means of a bar chart. The median response for these data is a rating of 3.5, the average of the two middle responses. Usually, however, it is more meaningful with data of this type to specify that the median response is between "fair" and "good" rather than to define it numerically as 3.5. The modal response for the data is the rating "good."

The question often arises as to the usefulness of reporting the mean of data measured on the ordinal scale. It is argued that since the intervals between units on the scale are not necessarily of equal size, the operations of arithmetic (addition, subtraction, etc.) are not valid. Hence, the arithmetic

Ordinal Scale

Table 1.9. Frequency table of quality of care ratings in geriatric care center

Rating	Frequency	Relative Frequency (%)
(1) Very poor	3	15
(2) Poor	3	15
(3) Fair	4	20
(4) Good	7	35
(5) Very good	3	15
	20	100

average of ordinal (or nominal) data is not considered, in general, to be a useful or valid descriptive measure. In spite of these arguments, however, it is common to see the mean reported for ordinal data. In such instances it has been assumed that the intervals of the scale are equal, or at least approximately so.

Both the interval and the ratio scale allow the use of ordinary arithmetic operations. Thus, for measurements in these scales, all of the descriptive statistics presented in this chapter may be employed. The choice of an appropriate measure of central tendency is influenced by the shape of the frequency distribution and the use to which the chosen summary will be put.

For symmetric distributions the mean and the median are equal. For symmetric or nearly symmetric sample data there is no distinction between the two. For skewed data, however, the mean and the median are not equal. The rationale for choosing between the mean and the median can best be illustrated by an example.

The distribution of "length of stay" in days among hospitalized patients is of interest to both the hospital administrator and the medical staff. The administrator must plan for the use of available facilities and the physicians must monitor the quality of care. Particularly difficult cases might require extended hospitalization. Although relatively few in number, such cases could skew the frequency distribution so that the mean and the median would not be equal.

The administrator, in planning the allocation of resources (e.g., dietary, housekeeping, laundry), would want to know how long a new patient might be "expected" to stay and consequently would use the mean length of stay as the statistic of choice.

The medical staff, in reviewing its performance and taking into account the fact that some lengthy hospitalizations are unavoidable, would want to know that the number (or proportion) of patients with extended stays was within acceptable limits. Thus, they might use the median to estimate the day by which 50 percent of the patients should be discharged.

When the frequency distribution of a set of measurements is highly skewed and the sample size is small, the median usually will provide a more meaningful representation of the center of the distribution.

Consider, for example, a study in which the data consist of survival time for patients receiving a particular regimen for treatment of an advanced stage of cancer of a specified site. The survival times in months for five patients are 3, 5, 4, 4, 29.

The mean ($\bar{Y} = 9$ months) as a measure of the "center" of the above set of data gives a misleading impression of the efficacy of the treatment since it is unduly influenced by the extreme value, or "outlier" 29 months. The median (4 months), on the other hand, since it divides a distribution by count, is more representative of the central tendency of the data. Note, however, that as the number of observations in the data set becomes larger, the influence on measure of location exerted by the presence of a few extreme values becomes less critical.

Often, no one descriptive measure is clearly better than the others. In this case, it may be of benefit for complete description to present a frequency table and/or a histogram, and the mean, median, mode, range, and standard deviation of the data.

PROBLEMS

1. In an experiment to determine the effect of a certain drug on serum cholesterol level (measured in mg/100 ml) in 30-year-old males, the following data were recorded for the drug treated group:

230	235	200	$n = 30$
175	170	290	
181	245	150	
190	120	145	
220	225	215	$\sum Y_i = 6311$
195	200	230	$\sum Y_i^2 = 1371111$
240	200	230	
165	265	210	
250	210	215	
190	270	250	

a. Using an interval width of 20, set up the frequency table, the relative frequency table, and the cumulative frequency table for the above data.

b. Draw the histogram. Be sure to label axes.

c. Calculate:
   i. mean   iv. range
   ii. median   v. variance
   iii. mode   vi. standard deviation

2. For the set of numbers (5, 3, 2, 1, 4), calculate the variance using the theoretical formula

$$s^2 = \frac{\sum_{i=1}^{n} (Y_i - \bar{Y})^2}{n - 1}$$

and the computational formula

$$s^2 = \frac{\sum_{i=1}^{n} Y_i^2 - \frac{(\sum Y_i)^2}{n}}{n - 1}$$

3. Consider the following data:

1	2	1	2
4	3	3	4
6	5	8	7
10	2	1	7
12	9	3	6

  a. Using intervals of width 2, draw the histogram.
  b. Is the distribution of data skewed or symmetric?
  c. Calculate the mean and the median. How do these statistics compare?

4. In a study to determine the effect of cigarette smoking on phenacetin metabolism, phenacetin level in plasma ($\mu g/ml$) was measured 2 hours after administration of the drug to 10 smokers and 12 nonsmokers. The results are given below.

Phenacetin ($\mu g/ml$)	Frequency Nonsmokers	Smokers
0.005–0.505	1	4
0.505–1.005	1	2
1.005–1.505	2	2
1.505–2.005	3	1
2.005–2.505	1	0
2.505–3.005	2	1
3.005–3.505	1	0
3.505–4.005	1	0
	12	10

Summary Statistics	Nonsmokers	Smokers
Mean	2.0	0.5
Standard dev.	0.7	0.3

  a. Make a relative frequency table and a cumulative frequency table for the above data.
  b. Why must relative frequency rather than absolute frequency be used when comparing the two groups?
  c. What percentage of the observations fall below the mean for the non-smoking group? For the smoking group?
  d. Is the distribution of the phenacetin concentrations for the smokers skewed or symmetrical?
  e. For the smoking group, which would be larger, the mean or the median?
  f. Based on the frequency tables, what can you say about the effect of smoking on phenacetin concentration?

5. In a clinical laboratory, tests were run on three new instruments used for making a certain blood chemistry measurement. Tests solutions were prepared containing a known concentration (10 mg/ml) of the substance to be measured. The following results were obtained with the three experimental instruments:

	Instrument	
1	*2*	*3*
5	10	10
10	9	11
7	10	9
15	9	10
16	11	10
12	8	9
4	9	11
8	7	12
10	8	8
13	9	10
$\sum Y$: 100	90	100
$\sum Y^2$: 1148	822	1012

a. Determine the mean and the standard deviation for the three instruments. In clinical measurements, three terms that are used frequently are precision, unbiasedness, and accuracy. *Precision* refers to the spread of a set of observations and is measured by the standard deviation. *Unbiasedness* refers to the tendency of a set of measurements to be equal to a "*true*" value. For an instrument to be *accurate*, its readings must be both precise and unbiased.

b. Describe all three instruments in terms of the above definitions.

c. Which instrument would you purchase? Why?

6. The following table presents data from two different hospital laboratories, all making the same specified blood chemistry determination. Within each lab, there are two technicians. Using a solution with a known concentration (5 mg/ml), each technician made three readings.

	Technician	
Lab	*1*	*2*
1	5, 8, 6	4, 4, 8
	[19, 125]	[16, 96]
2	8, 8, 4	5, 6, 6
	[20, 144]	[17, 97]

The numbers enclosed in brackets below each set are the sum and sum of squares, respectively, for each set of three readings.

a. Determine the mean and the variance for each technician. What factors might affect the variability within a given technician?

b. Based on the above information, which technician would you hire? Why?
c. Using the mean values for each technician as single observations, determine the mean and the variance for lab 1 and lab 2.
d. What factors might affect variability within a given lab?
e. Which lab produces the most accurate results?

CHAPTER 2

ELEMENTARY PROBABILITY

OVERVIEW

The study of elementary probability theory can be one of the most frustrating experiences a student can have. There are many reasons for such frustrations, especially if one attempts to establish a philosophical and mathematical basis for probability statements. In order to avoid problems concerning the "true" meaning of probability (such arguments still occupy professional mathematicians), we are going to take a very pragmatic approach.

Subjective concepts of probability seemingly play a large part in our daily lives. The words "probable" and "probably" are used frequently in ordinary conversation. Furthermore, by use of the synonym "likely" we can establish a scale of credence or faith on which to base our actions. Thus, an event of interest can be judged to be "unlikely," "likely," or "most likely" to occur. To this scale we must add, of course, "not certain" and "certain." This scale suffices for common discourse and for those situations in which our knowledge is limited. The problem that faces us as scientists is to establish a scale of measurement of uncertainty that has the same meaning for our colleagues as it does for us. Elementary concepts of probability provide the means for measuring the uncertainty of decisions and inferences based on data from samples and experiments.

OBJECTIVES

- To use frequency tables to answer questions about probability
- To find simple and compound probabilities by applying the addition and multiplication rules for combining probabilities
- To calculate conditional probabilities
- To apply Bayes' Rule to the solution of simple diagnostic problems

- To state the properties of a binomial experiment
- To apply the binomial formula to probability problems
- To select the appropriate technique for solving probability problems (e.g., Bayes', binomial, etc.)

FOUNDATIONS

Introduction

Usually, probability problems are approached through discussions about tossing coins, rolling dice, or selecting cards. These devices may be useful from a theoretical point of view but are not particularly instructive for students not majoring in statistics. We will use common biomedical situations for instruction in the assessment and use of measures of probability.

Table 2.1 shows a frequency table for serum cholesterol levels in normal men 40 to 59 years of age. The distribution is symmetric and unimodal with

Table 2.1. Grouped frequency distribution of serum cholesterol levels in normal men aged 40–59

Serum Cholesterol (mg/100 ml)	Frequency	Relative Frequency (%)	Cumulative Frequency (%)
119.5–139.5	10	1.0	1.0
139.5–159.5	21	2.0	3.0
159.5–179.5	37	3.5	6.5
179.5–199.5	97	9.3	15.8
199.5–219.5	152	14.5	30.3
219.5–239.5	206	19.7	50.0
239.5–259.5	195	18.6	68.6
259.5–279.5	131	12.5	81.1
279.5–299.5	96	9.2	90.3
299.5–319.5	47	4.5	94.8
319.5–339.5	30	2.9	97.7
339.5–359.5	13	1.2	98.9
359.5–379.5	6	0.6	99.5
379.5–399.5	4	0.4	99.9
399.5–419.5	—	0	99.9
419.5–439.5	1	0.1	100.0
439.5–459.5	—	0	100.0
459.5–479.5	1	0.1	100.1
Total	1047	100.1	

Adapted from *Circulation* 16:227, 1957.

the peak in the interval of 219.5–239.5 mg/100 ml. From the cumulative percent column we see that 50.0 percent of these normal men had cholesterol levels measuring less than 240 mg/100 ml serum.

Let us see how we can relate these data to probability. If you were to select a normal male at random from this group of 1047, what would be the probability that his serum cholesterol measurement was in the range of 160–179? From Table 2.1 we see that 37 of the 1047 have levels in this range. Clearly, the chance of selecting one of the 37 is (37/1047) × 100 = 3.5 percent as shown in the percent relative frequency column of the table. This demonstrates the simplest definition of probability: The probability P that a particular outcome will occur is the ratio of the number of times the outcome can possibly occur to the total possible number of outcomes.

In the example, the outcome is choosing a person whose level is in the range 160–179; the number of men with levels in this range is 37; and the total possible number of men who could be selected is 1047. Thus, $P = 37/1047 = 0.035$. For convenience we will use "percent chance" and probability interchangeably (e.g., 3.5 percent and 0.035). Thus, just as the percent relative frequency of the whole table add up to 100, the probabilities for all the class intervals would add up to 1.00. Hence any probability value will lie between 0 and 1.

Another and conceptually slightly different problem: if you were to select a normal male aged 40–59 at random from the general population from which these 1047 were drawn, what is the probability that his cholesterol value would be less than 200?

Your answer should be 15.8 percent. The conceptual problem here is whether this sample of 1047 normal males adequately represents the total population of normal men from which it was drawn. If it does, we can answer probability questions both about the sample and about the population that it represents using these data.

Calculating Probabilities

Addition of probabilities.   As we have seen, the probability of a single event can be calculated by enumerating the number of ways the outcome can occur and dividing by the total number of outcomes. This can be extended easily to calculating the probability of more than one event.

Referring to Table 2.1, what is the probability that a person drawn at random would have a cholesterol level either below 160 mg/100 ml or 340 mg/100 ml or greater? Clearly this either/or situation can be satisfied by drawing one of the 21 + 10 = 31 persons with levels below 160 mg/100 ml or one of the 13 + 6 + 4 + 1 + 1 = 25 persons with levels of 340 mg/100 ml or greater. Thus, the desired probability is given by $P = 31/1047 + 25/1047 = 0.030 + 0.024 = 0.054$. Alternatively, we can say there is a 5.4 percent chance that a person selected at random from among the 1047 will have a blood cholesterol level in one of the specified ranges.

Note that the two conditions specified cannot be satisfied at the same time by one person. That is, a person with a cholesterol level below 160 mg/100 ml cannot, at the same time, have a level greater than or equal to 340 mg/100 ml. When two or more conditions cannot be fulfilled at the same time they are said to be *mutually exclusive.*

When two events or outcomes are mutually exclusive, the probability that either one or the other will occur is given by the sum of the individual probabilities.

Symbolically, we may say that if two events, one denoted by A and the other by B, are mutually exclusive, then

$$P(A \text{ or } B) = P(A) + P(B)$$

This is read "The probability of either A occurring or B occurring is equal to the probability of A plus the probability of B."

In the foregoing example, A is the event that the person drawn at random would have a cholesterol level below 160 mg/100 ml, whereas B is the event that the person drawn at random would have a cholesterol level of 340 mg/100 ml or greater. Then

$$P(A) = 0.030, \qquad P(B) = 0.024$$

and

$$P(A \text{ or } B) = 0.030 + 0.024$$
$$= 0.054$$

When events are not mutually exclusive but can occur simultaneously, the probability of their joint occurrence must be taken into account.

The either/or statement of conditions in the preceding section is otherwise known as the "logical or." That is, one or the other of the conditions is satisfied but not both. To denote the simultaneous occurrence of two events the "logical and" is used. Thus, "A and B" means that both A and B occur.

The data in Table 2.2 will be used to illustrate the computation of probabilities where two events can occur jointly. Based on these data we may ask, "What is the probability that a person selected at random from the population represented by this sample will either be supported by public funds or have dental needs as his most urgent, or both?" The probability problem is compounded by the fact that one person could satisfy both conditions simultaneously. We must take this possibility into account.

Table 2.2. A survey of the relative urgency of dental and medical needs by source of financial support (hypothetical example)

Source of Funds	Most Urgent Need		Total
	Dental	Medical	
Public	470 B	280	750 A
Private	110	140	250
Total	580 A	420	1000

Let A be the event that the person selected is supported by public funds. Verify that $P(A) = 0.75$. Let B be the event that the most urgent needs of the person selected is dental. Verify that $P(B) = 0.58$.

Clearly, $P(A$ or $B)$ is not equal to the sum of $P(A)$ and $P(B)$ since this would be greater than 1. (Recall that probability values must lie between 0 and 1.)

The difficulty here is that the people who are both supported by public funds and have urgent dental needs have been counted twice. They were included once among those supported by public funds ($470 + 280 = 750$) and once among those who have urgent dental needs ($470 + 110 = 580$). Since the 470 people simultaneously satisfying both conditions have been counted twice, they must be subtracted once. The correct number of people in the sample who satisfy one or the other of the conditions, or both, is then given by

$$750 + 580 - 470 = 860 \quad = 470 + 280 + 110$$

Therefore,

$$P(A \text{ or } B) = 860/1000,$$
$$= 0.860$$

It is useful to rewrite this in extended form:

$$P(A \text{ or } B) = \frac{750 + 580 - 470}{1000}$$
$$= (750/1000) + (580/1000) - (470/1000)$$
$$= P(A) + P(B) - P(A \text{ and } B)$$

This is the most general form for the probability of either or both of two events. It is known as the *addition rule* of probability. It can be computed directly, as from Table 2.2, or indirectly if we know $P(A)$, $P(B)$, and $P(A \text{ and } B)$ but do not have a frequency table.

Conditional probability. Simple probability problems such as those discussed in the preceding section are based on selecting samples at random from the total population under study. Sometimes, however, we have additional information that will allow us to concentrate on only a part of a population. The part of the population under scrutiny usually has been characterized as that part having fulfilled a certain condition or conditions, hence the name *conditional probability*.

Looking again at Table 2.1, let us pose the question, "If a person selected at random from among the 1047 normal males is known to have a cholesterol level below 240 mg/100 ml, what is the probability that his true level is in the range of 120–139 mg/100 ml?"

From the table we see that $206 + 152 + 97 + 37 + 21 + 10 = 523$ persons satisfy the condition of having levels below 240 mg/100 ml. Moreover, 10 of these persons have levels in the range of 120–139 mg/100 ml. Thus, the required probability is $P = 10/523 = 0.019$.

Symbolically, let A denote the event that the person selected at random has a level below 240 mg/100 ml. Let B denote that the person also has a level in the range of 120–139 mg/100 ml. Then we have

$$P(B|A) = 10/523$$
$$= 0.019$$

which is read "The probability that the event B (level in the range of 120–139 mg/100 ml) will occur given that the event A (level below 240 mg/100 ml) has occured is equal to 0.019."

Clearly, all that has been done is to reduce the base population to those members who have satisfied the stated condition(s). Probabilities are then computed on this reduced population using the same procedures as before. This leads to an alternative formulation of conditional probability that can be used even when we cannot directly count the members of the population selected by the condition(s).

With event A and event B defined as above, it can be verified from Table 2.1 that

$$P(A) = 523/1047$$

and

$$P(A \text{ and } B) = 10/1047$$

Also,

$$P(B|A) = 10/523$$
$$= (10/1047)/(523/1047)$$
$$= \frac{P(A \text{ and } B)}{P(A)}$$

Thus, there are two ways to compute a conditional probability: by a direct count of the population or by knowing $P(A)$ and $P(A \text{ and } B)$. (Note that conditional probability has meaning only if $P(A) \neq 0$.)

Multiplication of probabilities. The formula defining conditional probability, $P(B|A) = P(A \text{ and } B)/P(A)$, can be rearranged by multiplying both sides by $P(A)$. Thus,

$$P(A \text{ and } B) = P(B|A)P(A)$$

In words, the probability that two events, A and B, both occur together is equal to the conditional probability that B occurs given that A occurred, multiplied by the probability that A in fact occurs.

This formula for the joint probability of two events is called the *multiplication rule* of probability.

Returning to the data in Table 2.2, let us see how the multiplication rule is applied. For a person selected at random from those in the sample, what is the probability that he is supported by public funds (event A) and his most urgent needs are dental (event B)?

From Table 2.2 we know directly that $P(A \text{ and } B) = 0.470$ so we can verify our calculations. Clearly,

$$P(B|A) = (470/750)$$

Also,

$$P(A) = (750/1000)$$

Thus, by the multiplication rule,

$$P(A \text{ and } B) = P(B|A)P(A)$$
$$= (470/750)(750/1000)$$
$$= (470/1000)$$
$$= 0.470$$

as we found by direct calculation.

It is important to note that there is a symmetry inherent in the multiplication rule. Clearly, the "logical and" does not depend on the order of the events. That is, A and B means exactly the same as B and A. Thus, we have

$$P(A \text{ and } B) = P(B \text{ and } A)$$
$$= P(A|B)P(B)$$

This merely indicates that the joint probability of two events can be computed in alternate but equivalent ways.

Independence of events.   If the occurrence of an event B is in no way affected by the occurrence of an event A, then the two events are said to be *independent*. Symbolically, if $P(B|A) = P(B)$, then A and B are independent.

The interested student can show that if $P(A) \neq 0$ and $P(B) \neq 0$, then $P(B|A) = P(B)$ implies $P(A|B) = P(A)$. That is, "B is independent of A" implies "A is independent of B."

The calculation of joint probabilities assumes a very simple form when the events are independent.

Let A and B be two independent events. Then,

$$P(A \text{ and } B) = P(A|B)P(B)$$
$$= P(A)P(B)$$

The probability that two independent events occur together is given by the product of their individual probabilities.

Returning to Table 2.1, if two men are selected at random out of the population from which the 1047 normal men were drawn, what is the probability that both men have serum cholesterol levels of less than 240 mg/100 ml?

Let A denote the event that the first person selected from the normal males aged 40–59 has a cholesterol level of less than 240 mg/100 ml. Verify that $P(A) = 0.500$. Let B denote the event that the second person selected has a cholesterol level of less than 240 mg/100 ml. Then $P(B) = 0.500$ also.

Since the selection of the first person from among the normal men is in no way connected with the selection of the second person (this is the meaning of independent random samples) and vice versa, we have $P(A|B) = P(A)$ and $P(B|A) = P(B)$. Then,

$$P(A \text{ and } B) = 0.500(0.500)$$
$$= 0.250$$

Bayes' Rule

$\partial M \xi T$

The relationships developed in the discussions in the preceding sections can be used to find the probability that one event will occur given that some other event has already occurred, or of more interest, the probability that one state or condition exists given that some other state or condition has been observed. Examples are the probability that a person has tuberculosis given that the skin test is positive, and the probability that a patient has a particular disease given that certain symptoms are present.

Suppose that the rate of a certain disease in a given population is 5 percent. That is, at any one time 5 percent of the total population are afflicted with the disease. Suppose also that 80 percent of those with the disease show certain symptoms while 10 percent of those without the disease also show the same symptoms. For a person selected at random from this population, what is the probability that the disease is present if the symptoms are observed? Symbolically, we need to evaluate $P(\text{disease}|\text{symptoms})$.

Let D be the event that the disease is present. Let $\bar{D}$ (read "not D") be the event that the disease is absent. Let S and $\bar{S}$ similarly stand for the presence and absence of the symptoms.

Verify that

$$P(D) = 0.05$$
$$P(\bar{D}) = 1 - P(D)$$
$$= 0.95$$
$$P(S|D) = 0.80$$
$$P(S|\bar{D}) = 0.10$$

We are interested in ascertaining the value of $P(D|S)$ using only the foregoing probability values. The utility of this approach is that patients are usually identified and information filed by disease rather than symptoms. Consequently, the probabilities involving D and $\bar{D}$ either directly or conditionally can be estimated from existing practice and methods. The probabilities conditional on symptoms are then inferred.

It may be instructive to note that most medical texts present diseases and describe the associated symptoms, whereas the practitioner will be faced with symptoms from which the disease is to be inferred.

Recalling the definition of conditional probability, we have

$$P(D|S) = \frac{P(D \text{ and } S)}{P(S)}$$

or, multiplying through by $P(S)$,

$$P(D \text{ and } S) = P(D|S)P(S)$$

Also,

$$P(S|D) = \frac{P(S \text{ and } D)}{P(D)}$$

or, multiplying through by $P(D)$,

$$P(S \text{ and } D) = P(S|D)P(D)$$

Now "D and S" and "S and D" mean exactly the same thing. Hence,

$$P(D|S)P(S) = P(S|D)P(D)$$

or

$$P(D|S) = \frac{P(S|D)P(D)}{P(S)}$$

From the data given in the problem we know the value of $P(S|D)$ and $P(D)$. What is the value of $P(S)$?

Clearly, anyone with the symptoms either has the disease or does not have the disease. Thus, each person can be classified as belonging to "S and D" or "S and $\bar{D}$." Note that these two classifications are mutually exclusive. Then we may write the probability of the event S (the presence of the symptoms) by using the addition rule of probability for mutually exclusive events:

$$P(S) = P(S \text{ and } D) + P(S \text{ and } \bar{D})$$

As before

$$P(S \text{ and } D) = P(S|D)P(D)$$

and

$$P(S \text{ and } \bar{D}) = P(S|\bar{D})P(\bar{D})$$

Therefore,

$$P(S) = P(S|D)P(D) + P(S|\bar{D})P(\bar{D})$$

Hence,

$$P(D|S) = \frac{P(S|D)P(D)}{P(S)}$$

$$= \frac{P(S|D)P(D)}{P(S|D)P(D) + P(S|\bar{D})P(\bar{D})}$$

This is known as Bayes' Rule for computing one probability from the knowledge of others.

The necessary values are all given in, or can be calculated from, the statement of the problem:

$$P(D|S) = \frac{0.80(0.05)}{0.80(0.05) + 0.10(0.95)}$$

$$= 0.30 \text{ pts w7 disease who express symptoms}$$

That is, if a person is observed to have the symptoms, there is 30 percent chance that he has the disease.

In the context of a screening examination, this result is interpreted as follows. The frequency of the disease in the population is 5 percent so the probability for any person who walks in the door to have the disease is $P(D) = 0.05$. That is, one in 20 of the people screened are expected to have the disease. A particular person is examined and found to have the symptoms. The probability that he has the disease is then increased to 30 percent, or $P(D|S) = 0.30$. The presence or absence of the disease will be confirmed by subsequent definitive tests.

This use of Bayes' Rule is based on the concept that probabilities of events should depend on the amount of information that we have. For example, before a physician sees a waiting patient, all disease states are possible. Then, as the patient is inspected visually, certain disease states are eliminated (probability approaches zero) and other diseases may appear more likely (probability increases). Subsequent information, in the form of history, physical, and laboratory findings, then lead to the most probable diagnosis. The mathematical formalization of this process is contained in the development of Bayes' Rule.

A more pragmatic approach to Bayes' Rule is to recognize that it is really nothing more than an algebraic relationship that can be used to compute probabilities of interest. This algebraic relationship is simply a consequence of the rules governing the calculation of probabilities. The difficulty with using Bayes' Rule as a basis of differential diagnosis is that good estimates of the required probabilities are usually not available. Despite this drawback, however, there is much interest at present in the development of diagnostic models based on these ideas.

Binomial Experiments

In its classic form the description of a scientific experiment includes a section on methods and materials which describes the procedures, the materials used, and the conditions under which the experiment was carried out. With this information other workers can repeat the experiment independently using the same materials under the same conditions. Random sampling is an example of repeated independent experiments. Each member of the sample is independent of every other member and the observation of each member represents a "trial" of the experiment.

Many applications of probability can be reduced to the model of independent repeated trials of an experiment for which there are only two possible outcomes at each trial. When the experiment has only two possible outcomes and the probability of each outcome remains constant over repeated independent trials, it is called a *binomial experiment*. The traditional model for the binomial experiment is the tossing of a coin with the resulting display of heads or tails.

The properties of a binomial experiment are summarized as follows:

1. There must be a definite number of trials
2. Each trial must result in one of two possible outcomes
3. The probabilities assigned to each of the two outcomes must be constant from trial to trial
4. Each of the trials must be independent of the others

Usually, since only two outcomes are possible, we denote the probability of the event of interest by p and the other by $q = 1 - p$. Examples include: the transmission of a genetic disorder, a student passing or failing, a delivery resulting in a male or female infant, a patient requiring treatment or no treatment, a patient having a positive or negative skin reaction.

Suppose we wish to investigate the presence of a certain genetic disorder among three siblings. All possible outcomes are listed in Table 2.3 where

Table 2.3. Possible outcomes for the transmission of a genetic disorder to three siblings (N = normal, D = defective)

Outcome for				
Sibling 1	Sibling 2	Sibling 3	Probability	No. of Normals
N	N	N	$p \cdot p \cdot p$	3
N	N	D	$p \cdot p \cdot q$	2
N	D	N	$p \cdot q \cdot p$	2
D	N	N	$q \cdot p \cdot p$	2
N	D	D	$p \cdot q \cdot q$	1
D	N	D	$q \cdot p \cdot q$	1
D	D	N	$q \cdot q \cdot p$	1
D	D	D	$q \cdot q \cdot q$	0

sibling 1 is the first born, 2 is the second born, and so forth, and p is the probability of a normal child. Then $q = 1 - p$ is the probability of a child with the genetic disorder.

The probabilities in Table 2.3 have been computed by extending the concept of multiplying probabilities of two independent events to more than two events. Thus, the probability that sibling 1 is normal *and* sibling 2 is normal *and* sibling 3 is normal is given by the product of the independent probabilities.

It is legitimate to multiply probabilities in this fashion for trials in a binomial experiment.

A summary of the outcomes, neglecting the order of the children, is shown in Table 2.4. From the listing of all possible sequences it is seen that

Table 2.4. Summary of the possible outcomes for the transmission of a genetic disorder to three siblings in terms of the number of normal children

No. of Normals (k)	Number of Ways k Normals Can Occur	Probability of k Normals
0	1	q^3
1	3	$3pq^2$
2	3	$3p^2q$
3	1	p^3

there is only one way for none of the siblings to be normal, three ways for there to be one normal child, three ways for there to be two normal children, and one way for all the siblings to be normal.

The probabilities shown in the last column can be justified by invoking the addition rule for probabilities. Let the genetic outcome in order of birth be denoted as follows: NDN means first sibling is normal, second sibling is defective, third sibling is normal, etc.

What is the probability that among the three siblings there will be exactly two normal children? That is, what is the value of P(NND or NDN or DNN)?

Since no two of these events can occur together (i.e., if one birth order exists then the others do not), we simply add the appropriate probabilities, Thus,

$$P(\text{NND or NDN or DNN}) = P(\text{NND}) + P(\text{NDN}) + P(\text{DNN})$$

$$= p \cdot p \cdot q + p \cdot q \cdot p + q \cdot p \cdot p$$

$$= 3p^2q$$

Note that we have invoked the multiplication rule within each sequence of births since NND is just a shorthand notation for "N and N and D" in the specified order.

You should verify the other probabilities listed in Table 2.4.

Another useful way to visualize binomial experiments is to draw "tree diagrams." One tree starts with one of the outcomes at the top and the other tree starts with the second outcome. The branches of the tree will then lead to one or the other of the two outcomes and then branch further. The branches are then labeled with the appropriate probabilities.

The problem on the genetic disorder among three siblings illustrates the technique.

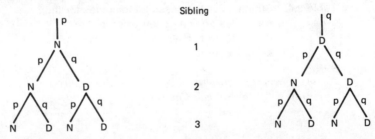

The probability of any sequence is found by multiplying the probabilities encountered as that branch is traced. For example, the sequence corresponding to no normal births (i.e., $k = 0$) is represented by the rightmost path in the tree on the right:

$$P(\text{No. normals} = 0) = q \cdot q \cdot q$$
$$= q^3$$

Similarly, other particular sequences can be enumerated. For example, "only one normal" is satisfied by the individual branches shown and

$$P(\text{No. normals} = 1) = pqq + qpq + qqp$$
$$= 3pq^2$$

as before.

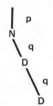
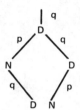

Mathematically, the process outlined above leads to what is termed the binomial probability generating function.

Specifically, for the genetic example in Table 2.3 we may write

$$(p + q)^3 = (p + q) \cdot (p + q) \cdot (p + q)$$
$$= (p^2 + 2pq + q^2) \cdot (p + q)$$
$$= p^3 + 3p^2q + 3pq^2 + q^3$$

Note that each of the terms in this expression appears in the column of probabilities in Table 2.4 where

$p = P(\text{normal})$

$q = P(\text{defective})$

$3 = \text{Number of siblings}$

To compute any probability, simply select the term for which the exponent of p is equal to the number of normal siblings of interest (recall that $x^0 = 1$ for any number x, so that $p^3 = p^3 q^0$ and $q^3 = p^0 q^3$):

$$P(0 \text{ normal}) = q^3$$
$$P(1 \text{ normal}) = 3pq^2$$
$$P(2 \text{ normals}) = 3p^2 q$$
$$P(3 \text{ normals}) = p^3$$

In general, for any binomal experiment, with outcome A or $\bar{A}$ at any trial,

$$(p + q)^n = \binom{n}{0} p^0 q^{n-0} + \binom{n}{1} p^1 q^{n-1}$$
$$+ \cdots + \binom{n}{m} p^m q^{n-m} + \cdots + \binom{n}{n} p^n q^0$$
$$= \sum_{k=0}^{n} \binom{n}{k} p^k q^{n-k}$$

where

$$p = \text{probability of } A \text{ at any trial,}$$
$$q = 1 - p$$
$$n = \text{number of trials,}$$
$$\binom{n}{m} = \frac{n!}{m!(n-m)!}$$

with $n!$, read n—*factorial*, given by

$$n! = n(n-1)(n-2) \cdots (2)(1)$$

Also,

$$m! = m(m-1)(m-2) \cdots (2)(1)$$
$$(n-m)! = (n-m)(n-m-1) \cdots (2)(1)$$

and

   $0! = 1$ by definition

Each term on the right side of the binomial generating equation corresponds to a specified set of paths down the appropriate tree diagrams.

   Let us see how the binomial probability generating function, usually called simply the "binomial" can be applied to the probability of a given number of genetic disorders among three siblings as discussed above. Thus, with $n = 3$ and $P(\text{normal}) = p$, $P(\text{defective}) = q$,

$$\sum_{k=0}^{3} \binom{3}{k} p^k q^{3-k}$$

$$= \binom{3}{0} p^0 q^{3-0} + \binom{3}{1} p^1 q^{3-1} + \binom{3}{2} p^2 q^{3-2} + \binom{3}{3} p^3 q^{3-3}$$

$$= \frac{3!}{0!(3-0)!} p^0 q^3 + \frac{3!}{1!(3-1)!} p^1 q^2$$

$$+ \frac{3!}{2!(3-2)!} p^2 q^1 + \frac{3!}{3!(3-3)!} p^3 q^0$$

$$= \frac{3 \cdot 2 \cdot 1}{1 \cdot (3 \cdot 2 \cdot 1)} p^0 q^3 + \frac{3 \cdot 2 \cdot 1}{1(2 \cdot 1)} p^1 q^2$$

$$+ \frac{3 \cdot 2 \cdot 1}{2 \cdot 1(1)} p^2 q^1 + \frac{3 \cdot 2 \cdot 1}{3 \cdot 2 \cdot 1(1)} p^3 q^0$$

$$= p^0 q^3 + 3p^1 q^2 + 3p^2 q^1 + p^3 q^0$$

$$= q^3 + 3pq^2 + 3p^2 q + p^3 \text{ as before}$$

   A further refinement in notation makes the use of the binomial much more productive. Suppose that the outcome of interest in a binomial experiment is symbolized by the letter Y. (Note that we need label one outcome only since if it did not occur then the other outcome must have occurred.) Then we may write

$$P(Y = k \mid n, p) = \binom{n}{k} p^k q^{n-k}, \quad k \le n$$

which equals the probability that the outcome Y occurs k times in n trials when the probability of the occurrence of Y in a single trial is p. Thus, the general rule for the sibling example is

$$P(\text{No. normals} = k|3, p) = \binom{3}{k} p^k q^{3-k}, \qquad k = 1, 2, 3$$

The binomial probability function also can be used to calculate probabilities of compound events. Recall that if events are mutually exclusive then the probability of the occurrence of any of the events is equal to the sum of the individual probabilities. Thus,

$$P(\text{two or more normals}|3, p) = P(k = 2|3, p) + P(k = 3|3, p)$$

$$= \binom{3}{2} p^2 q + \binom{3}{3} p^3 q^0$$

$$= 3p^2 q + p^3$$

We could also write

$$P(\text{two or more normals}|3, p) = \sum_{k=2}^{3} \binom{3}{k} p^k q^{3-k}$$

In general,

$$P(Y \le m|n, p) = \sum_{k=0}^{m} \binom{n}{k} p^k q^{n-k} \qquad m \le n$$

$$P(Y \ge m|n, p) = \sum_{k=m}^{n} \binom{n}{k} p^k q^{n-k} \qquad m \le n$$

Suppose that the probability of transmitting a defective gene is 1/4. Verify that for a family of four children the probability of one normal child or no normal children is 13/256; that the probability of exactly two defective children is 27/128.

METHODOLOGY

Frequency Tables

Table 2.1 is an empirical frequency table that can be used to represent the population from which it was drawn. From the table we can find the upper cholesterol value that delimits approximately 95 percent of normal men. This value is 320 mg/100 ml (actually 5.2 percent lie above 320 mg/100 ml so this value marks off 94.8 percent of the sample). We can interpret this to say that there is approximately a 95 percent chance that a normal male aged 40–59 selected at random will have a cholesterol value below 320 mg/100 ml. What would be the chance that a man selected at random would have a serum cholesterol equal to or greater than this value?

Clearly, this question can be answered directly from the table. We see that 94.8 percent of the values are lower than 320 mg/100 ml. Consequently, $100\% - 94.8\% = 5.2\%$ of the values are higher.

What percentage of the sample of normal males in Table 2.1 is between 180 mg/100 ml and 320 mg/100 ml?

We have seen above that 5.2 percent of the measurements are 320 mg/100 ml or higher. Verify that 6.5 percent are below 180 mg/100 ml. Thus, $100\% - 5.2\% - 6.5\% = 88.3\%$ are between the two stated limits.

Then it can be stated that we are 88.3 percent *confident* that a normal male aged 40–59 drawn at random from this population would have a cholesterol level in the range 180–320 mg/100 ml.

Randomization Problems

A researcher has four rats to use in an experiment. Unknown to him, two of the rats are in the early stages of an illness. This could bias his results since the performance of a sick rat is different from that of a well rat. The researcher randomly assigns two rats to the experimental treatment. The remaining two will act as controls. What is the probability that the two sick rats will both be assigned to the experimental group?

Often the best way to begin on small problems of this nature is to list the possible outcomes and then enumerate the cases of interest. Assume the rats are identified with the labels 1, 2, 3, 4, with 1 and 2 being the sick rats.

Possible Samples with Two Rats/Group

Experimental	Control
1, 2	3, 4
1, 3	2, 4
1, 4	2, 3
2, 3	1, 4
2, 4	1, 3
3, 4	1, 2

Thus, we see that there are six possible ways to arrange the four rats into two groups of size two. Also, only one arrangement puts both rat 1 and rat 2 into the experimental group. Therefore, the probability that both sick rats will be in the experimental group is 1/6.

Similarly, what is the probability that only one of the sick rats will be assigned to the experimental group? Clearly, four of the six cases satisfy the condition so that the probability is $4/6 = 2/3$.

It is instructive to note that the possible arrangements of the four rats favor neither the experimental nor the control group. That is, the various ways that the sick animals can be included into the control group is just the mirror image of the ways to include them in the experimental group. Moreover, there are more cases where the sick and the well animals are mixed than there are for all one kind to be selected into one group. As you can imagine, when the group size increases, even if one-half of the subjects are somehow not like the other half, the probability that one group would have all or nearly all one kind of subject would be extremely small. The function of randomization is then seen to be to spread out any chance differences into all the study groups.

Two-way Tables

A study was conducted to determine the effectiveness of a particular test for detecting a certain disease. Of the 100 people randomly selected, the standard method showed that 10 had the disease and 90 did not. In the nondisease group 86 individuals had negative tests and 4 had positive tests. The disease group had 3 individuals with negative tests and 7 with positive tests.

These outcomes may be represented in the following form:

	Test		
	−	+	Totals
No disease	86	4	90
Disease	3	7	10
Totals	89	11	100

Using the information above, what is the probability that an individual selected at random from the population would have the disease? To answer this we must use the "marginal" totals. Of the 100 individuals in the study, 10 have the disease. Therefore,

$$P(\text{disease}) = \frac{10}{100} = 0.10$$

Similarly,

$$P(\text{positive test}) = \frac{11}{100} = 0.11$$

The conditional probability that a person with a positive test would have the disease is

$$P = (\text{disease}|\text{positive test}) = \frac{7}{11} = 0.64$$

This can also be determined as

$$P(\text{disease}|\text{positive test}) = \frac{P(\text{disease and positive test})}{P(\text{positive test})}$$

$$= (7/100)/(11/100)$$

$$= 7/11$$

as before.

Verify the following:

$$P(\text{negative test}|\text{disease}) = 0.30$$

$$P(\text{positive test}|\text{no disease}) = 0.044$$

Bayes' Rule

A group of researchers wanted to establish the probability of 5 or more years' survival for women who had some spread of cancerous tissue to the lymph glands from an involved breast. A retrospective study was done by identifying the records of women who had undergone a radical mastectomy and reviewing these records for 5 or more years' survival and the presence of one to three lymph nodes showing cancer cells.

It was found that the frequency of 5-year survival for the group studied was 90 percent. The frequency of one to three involved lymph nodes among the 5-year survivors was 17 percent. The frequency of finding a record of involved lymph nodes among the women who did not survive 5 years was 72 percent.

What is the chance of a woman's surviving 5 years following radical mastectomy if it is found that one to three lymph nodes are involved?

Let A be the event that one to three involved lymph nodes are found. Let B be the event that survival is five years or greater.

Then, using $\bar{A}$ and $\bar{B}$ to mean "not A" and "not B," respectively,

$$P(A|B) = 0.17, \qquad P(A|\bar{B}) = 0.72$$

$$P(B) = 0.90, \qquad P(\bar{B}) = 0.10$$

Table 2.5. Results of a screening test applied to diseased and nondiseased subjects in a random sample from the population

Test Result	Disease Present	Disease Absent	
Positive	A	B	$A + B$
Negative	C	D	$C + D$
	$A + C$	$B + D$	$A + B + C + D$

Thus,

$$P(B|A) = \frac{P(A|B)P(B)}{P(A|B)P(B) + P(A|\bar{B})P(\bar{B})}$$

$$= \frac{0.17(0.90)}{0.17(0.90) + 0.72(0.10)}$$

$$= 0.68$$

There is a 68 percent chance of surviving 5 years or more if one to three nodes are found.

Bayes' Rule can also be used to assess the predictive value of a screening test. The purpose of a screening test is to identify, through a reasonably cheap, fast, and safe method, individuals who currently have or are at an increased risk for developing a disease. At the same time, the screening test should exclude members of the population who are disease-free or who are not at increased risk.

The preferred method for assessing the value of a screening test is to apply it to a group of subjects selected at random from the appropriate population. Each subject is classified as being either positive or negative according to the test criterion (e.g., positive for diabetes if the blood glucose exceeds a certain value). Using standard criteria the subjects are then further classified into a group known to have the disease and into a group of subjects known not to have the disease. This classification of subjects must be free of error and not dependent on the screening test. The data from such a process would be displayed as in Table 2.5. From the $A + C$ diseased subjects, A were *true positives* and C were *false negatives* according to the test.

From the $B + D$ nondiseased subjects, B were *false positives* and D were *true negatives* according to the test.

For a perfect test (i.e., no misclassifications at all) we would have $C = 0$ and $B = 0$. Of course, this can never be obtained in practice.

Table 2.5 yields the following important definitions:

Sensitivity of the test = positivity in disease

$$= A/(A + C)$$

Specificity of the test = negativity in health

$$= D/(B + D)$$

Prevalence* of the disease = $(A + C)/(A + B + C + D)$

Predictive value of the test = $A/(A + B)$

We shall now show how the predictive value of the test is related to Bayes' Theorem. Let "+" and "−" indicate positive and negative results respectively. Also, let D and $\bar{D}$ indicate respectively the presence and absence of the disease. Then, from Bayes' Theorem we may write the probability that an individual has the disease given a positive result on the screening test as

$$P(D|+) = \frac{P(+|D)P(D)}{P(+|D)P(D) + P(+|\bar{D})P(\bar{D})}$$

Now,

$$P(+|D) = A/(A + C)$$

$$P(+|\bar{D}) = B/(B + D)$$

$$P(D) = (A + C)/(A + B + C + D)$$ Is D true neg. or ⊕ Disease?

Disease

and

$$P(\bar{D}) = (B + D)/(A + B + C + D)$$

whence,

$$P(+|D)P(D) = \frac{A}{A + C} \cdot \frac{(A + C)}{(A + B + C + D)}$$

$$= \frac{A}{A + B + C + D}$$

* From the description of Table 2.5 this is actually the *point prevalence* of the disease. See Chapter 9 for further discussion.

$$P(+|\bar{D})P(\bar{D}) = \frac{B}{(B + D)} \cdot \frac{(B + D)}{(A + B + C + D)}$$

$$= \frac{B}{A + B + C + D}$$

Therefore,

$$P(D|+) = \frac{\dfrac{A}{A + B + C + D}}{\dfrac{A}{A + B + C + D} + \dfrac{B}{A + B + C + D}}$$

$$= \frac{A}{A + B}$$

which we can verify directly from Table 2.5. Thus, $P(D|+)$ gives the predictive value of the test for a random sample from the population, as was stated.

Recall that if we select a subject at random from the population, the probability that he or she has the disease is given by $P(D)$, the prevalence of the disease in the population. Thus, the probability that the individual will have the disease is increased by added information that the screening test is positive.

When a random sample from the population is collected and subjected to the screen and subsequently diagnosed by standard criteria as above, all of the items of interest can be computed directly from the data and Bayes' Theorem is not needed. However, Bayes' Theorem can be used when the prevalence of a disease in the population is known and the sensitivity and specificity of a test are determined by applying it to a nonrandom series. For example, suppose a potential screening test is developed in a hospital setting where diseased persons are routinely diagnosed and nondiseased persons are readily available. Then $P(+|D)$ and $P(+|\bar{D})$ are ascertained on hospital patients and $P(D)$ and $P(\bar{D}) = 1 - P(D)$ are known from other sources. These values can then be used in Bayes' Theorem to determine the utility of the screening test in terms of its predictive value.

The predictive value of the test can be put directly into terms of sensitivity, specificity, and prevalence.

From Table 2.5 we see

$$\text{Sensitivity} = A/(A + C)$$

$$= P(+|D)$$

$$= \frac{\text{True positives}}{\text{True positives} + \text{false negatives}}$$

$$\text{Specificity} = D/(B + D)$$

$$= 1 - P(+|\bar{D})$$

$$= \frac{\text{True negatives}}{\text{True negatives} + \text{false positives}}$$

Therefore, from the nonrandom hospital series we can determine sensitivity and specificity. Finally, we see that

$$P(D|+) = \frac{\text{Sensitivity} \times \text{prevalence}}{\text{Sensitivity} \times \text{prevalence} + (1 - \text{specificity})(1 - \text{prevalence})}$$

$$= \frac{1}{1 + \dfrac{(1 - \text{specificity})(1 - \text{prevalence})}{(\text{sensitivity})(\text{prevalence})}}$$

Thus, to assess the predictive ability of a test, we need its sensitivity and specificity and the prevalence of the disease in the population.

From this formulation, there is one important conclusion. For given sensitivity and specificity, predictive ability decreases with decreasing prevalence. This becomes very important for rare diseases even for highly sensitive and specific tests. When sensitivity and specificity are determined from nonrandom series, great care should be exercised in determining the effect of prevalence on the value of the screening test.

Binomial Experiments

About 10 days after attending a children's party it was discovered that one of the party-goers came down with the measles. Assume that the probability for contracting the measles following exposure is $p = 0.05$ and that the chance of any child becoming infected at the party is independent of the chance of infection for any other child.

Among three siblings who had never had the measles what is the probability that at least one of them will develop the measles from exposure at the party?

Clearly, the condition will be satisfied if one, two, or three of the children develop the measles. Using the binomial formula,

$$P(1 \text{ or more cases}) = \sum_{k=1}^{3} \binom{3}{k} p^k q^{n-k}$$

$$= \binom{3}{1} pq^2 + \binom{3}{2} p^2 q + \binom{3}{3} p^3$$

$$= 3pq^2 + 3p^2 q + p^3$$

$$= P(1 \text{ case}) + P(2 \text{ cases}) + P(3 \text{ cases})$$

Thus,

$$P(1 \text{ or more cases}) = 3(0.05)(0.95)^2 + 3(0.05)^2(0.95) + (0.05)^3$$
$$= 0.14$$

for a 14 percent chance for one or more of the children to develop the measles.

PROBLEMS

1. You are a part of a group of 12 students with chronic colds who have volunteered to take part in a study of a new product for symptomatic relief. The strategy is to pass out, completely at random, to subgroups of four students, identically appearing preparations of the new product, a standard product, and a placebo (an inert preparation). Thus, each subject will get one preparation but will not know which one. Answer the following questions:
   a. What is the probability that you would get the new product?
   b. What is the probability that you will get one of the chemically active compounds?
   c. If your symptoms were cleared up after you took the preparation you received, could you say with 100 percent probability that the relief was caused by one of the active ingredients?

2. Suppose you were at a large hospital and wanted to do a survey on infant deaths, excluding neonatal deaths ("neonatal" refers to the first 4 weeks of life). The hospital keeps all the death records for infants under one year of age in a central file. The file is not in any particular order with respect to the purpose of your survey. Thus, you would select a record from the file, look at it to determine the age, keep it if the age was greater than 28 days and return it if not.

   If 72 percent of the infant deaths are classified as neonatal and if it takes approximately 2 minutes to select and check a record, how long would you *expect* to spend to collect 100 records for your study? How many records would you expect to check?

3. The failure rate for a cardiac arrest alarm in an intensive care bed is 1 in 1000. For safety, a duplicate alarm is installed. What is the probability that a cardiac arrest will not be signalled?

4. The tree structure can be used to answer questions about the probability of transmission of alleles, or combinations of alleles, in genetics problems. Recall that a gene pair is indicated by two letters written together: for example, Aa, AA, aa. This means that, relative to this gene, all sperm are either A or a and all eggs are either A or a. Suppose both parents

are Aa. Then we can construct the following tree diagram for genetic combinations from these particular parents: Now, by assigning probabilities to each of the branches of the tree, we can compute the probability of any particular combination or set of combinations. Usually a

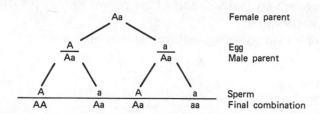

probability of 1/2 is assigned to each branch. Using this convention, what is the probability that the mother will pass her A on to one of her offspring? Clearly, this probability is given by $(1/2 \times 1/2) + (1/2 \times 1/2) = 1/2$ by following the two leftmost branches. More complex trees could be built describing parents starting with different gene combinations. Also, different generations and collateral lines could be diagrammed.

Similar trees can be constructed to describe the transmission of multiple genes from a particular parent. For example, suppose a man has the combination Aa and Bb located on different chromosomes. What is the probability that he will give his child the combination Ab? (i.e., the child will have AU and bV, where U and V are unspecified genes from his mother). Construct a tree similar to the one above to show that the path is Aa → A/Bb → b and the probability of Ab is $1/2 \times 1/2 = 1/4$. You will recognize that these problems really involve only meticulous bookkeeping. If you write down carefully the structure of the problem, then the solution should follow.

5. Assume the probability of the birth of a boy is one-half. What are the probabilities of the following sequences of single births?
   a. A succession of four boys?
   b. Two of each sex born alternately?

6. A man received from his father the following genes, all located in different chromosomes: A, B, C, and from his mother, a, b, c.
   a. What is the probability of his giving to his child the combination A, b, c?
   b. Either the combination A, B, C or a, b, c?

7. What is the probability that a man will transmit a specific allele to his granddaughter?

8. Assume a certain desirable trait of an individual depends on the presence of the following rare genes: A, B in heterozygous combination with their alleles, a, b. What is the probability that a grandchild of this individual will receive the alleles A, B from this grandparent?

9. A particular condition is present in 80 percent of the population. What is the probability that a sample of five people will <u>not</u> contain a person with the condition?

10. Data on a food poisoning outbreak are shown below.

	Ill	Not Ill	Total
Ate gravy	75	25	100
Did not eat gravy	5	10	15
Total	80	35	115

Calculate:
a. The probability of becoming ill after consuming the gravy
b. The probability of becoming ill if no gravy was eaten
c. The probability of not getting ill after consuming gravy

11. A certain clinical test is positive in 93 percent of the cases when the patient has the indicated disease. However, it is also positive in 3 percent of the cases when the patient does not have the disease. If the disease occurs in 1 percent of the population, what is the probability that a person selected at random and who has a positive test will also have the disease? Calculate the prevalence of the disease and the sensitivity, specificity, and predictive ability of the test.

12. A screening test for a rare form of veneral disease (VD) has a 7 percent false positive rate (i.e., positive test results for persons not having VD) and a 18 percent false negative rate (i.e., negative test results for persons with VD). In a population that has a 3 percent disease rate, what is the chance that someone who has a positive test also has this form of VD? What is the chance that someone with a negative test result has the disease? Calculate the prevalence of the disease and the sensitivity, specificity, and predictive ability of the test.

CHAPTER 3

POPULATIONS, SAMPLES, AND THE NORMAL DISTRIBUTION

OVERVIEW

The methods presented in the preceding chapters were concerned with the description of sample data and the basic principles of probability. In this chapter the concepts of frequency distributions, central tendency, variation, and probability will be used to develop the basis for inferential statistical methods. Chapters 4 through 7 will present methods for making statements about populations based on information obtained from samples. Chapter 3 thus provides a bridge between descriptive and inferential statistics.

The relationship between samples and populations and the properties of distributions of sample values in terms of the population values will be the starting point. Two very important frequency distributions, the normal distribution and the student-t distribution will be introduced and used to make statistical inferences about population values.

OBJECTIVES

- To define population and sample and state the relationship between them
- To describe the relationship between parameters and statistics and give examples of each
- To describe in words the differences between random and systematic samples
- To state the difference between empirical and theoretical distributions

- To list the properties of the normal distribution
- To state and interpret the Central Limit Theorem
- To find probabilities associated with the normal curve by using the normal table
- To relate the probabilities found using the normal table to the observed frequency table
- To place confidence intervals on the true population mean using the normal and student-t distributions

FOUNDATIONS

Populations and Samples

Fundamental to statistical analysis is the relationship between samples and populations. All the techniques presented in the remainder of this text are merely different methods for obtaining information about a population based on information contained in a sample from the population.

A question that arises is "why study a sample rather than the entire population?" Among the reasons why measuring every element in the population (i.e., performing a census) is not feasible are

1. The size of the population often makes it impossible or impractical to study in its entirety.
2. The cost of making observations on all elements of the population may be prohibitive.
3. All the individual members of the population may not be observable.
4. The measurement may be destructive. For example, if one wishes to determine the functional life of a heart pacer, the instrument must be tested until failure. Hence, the testing procedure is destructive.

A sample is not primarily interesting for its own sake but rather for what it reveals about the population. For example, a physician may wish to investigate the efficacy of a new treatment for gastric ulcerations. He selects 50 patients with the illness and administers the treatment. From the point of view of the research, the 50 patients do not constitute the population of interest. The physician wants to infer the effectiveness of the given treatment for *any* patient with the disease based on the results obtained from his sample of 50 patients. The statistical relationship between the population and the sample will allow the physician to make the desired inferences in a scientifically objective manner.

When numerical descriptive measures (e.g., mean and standard deviation) are calculated from a sample, they are referred to as *sample statistics*; when calculated using the entire population they are called *population parameters*. For clarity, sample statistics are denoted by English letters and population parameters are designated by Greek letters. The sample mean $\overline{Y}$ is an estimate of the population mean μ; the sample standard deviation s is an estimate of the population standard deviation σ.

In general, sample statistics will not be exactly equal to the corresponding parameters of the population from which the sample was drawn. For example, the mean $(\overline{Y})$ and standard deviation (s) of a random sample rarely, if ever, will be exactly equal to the true mean (μ) and the true standard deviation (σ) of the population. However, if we repeatedly select random samples from the population the sample statistics will cluster around the actual population parameters. Thus, the values of $\overline{Y}$ will cluster around the unknown value of μ. Similarly, the values of s will cluster around σ.

The objective in choosing a sample is to obtain observations that are representative of the parent population. The simplest and best known way to do this is to collect a *simple random sample* which is a sample selected in such a way that each observation or unit in the population has an *equal chance* of being included in any sample. It is important to note that randomness in the statistical sense does not mean haphazardness. To prevent either conscious or unconscious bias on the part of the investigator, the selection of a truly random sample can be accomplished only by mechanical means similar to drawing names out of a hat. Any particular simple random sample may not be a "minipopulation," but overall it is the best procedure to use to get representative samples.

Another widely used sampling technique is to select the sample according to some nonrandom, predefined system. An example of this type is the *systematic* sample. Interviewing every third patient on the daily hospital admissions list would be a systematic sample. With this type of sample, as with other nonrandom selection schemes, the possibility of obtaining a biased or nonrepresentative sample exists. A sample of patients located in even-numbered rooms may be affected by a systematic relationship to some extraneous variable. As the result of hospital construction or room numbering scheme, the even-numbered rooms may be higher priced, have easier access to nursing stations, have windows, and so forth, each of which may possibly affect patient response and thus bias the results of the study. It is extremely important that the investigator be aware of the possibility of such bias and prevent it whenever possible. Furthermore, any critical evaluation of a research study must include a scrutiny of the method of sample selection and an evaluation of the effect of possible sources of sample bias on the results of the study.

Theoretical Versus Empirical Distribution

A frequency distribution tabulated from sample data is called an *empirical* distribution and is used to approximate the true distribution of the population. By contrast, the frequency distribution of *all* the values in the population would be the underlying *theoretical* distribution.

When the purpose of a study is simply to describe the sample at hand, numerical summaries based on the empirical distribution suffice. However, when the purpose is to make inferences about the characteristics of the parent population from which the sample is drawn, the underlying theoretical distribution of all population values is used. Ordinarily, from prior knowledge of the characteristics of the population under study, the appropriate theoretical distribution is known. The *binomial* distribution was used in the preceding chapter. In this chapter the *normal* distribution will be emphasized. Other theoretical distributions will be presented later in the text.

The underlying theoretical distribution depends on the nature of the variable being observed. A variable that is measured in an experiment or collected in a sample is called a random variable (usually denoted by Y) and may be classified as either *discrete* or *continuous*.

Variables whose measurements are integers resulting from a count of some sort are called discrete variables. Examples include the number of bacteria observed on a series of slides, the number of children per family, and the number of diseased patients by economic status.

A continuous variable, on the other hand, may assume any value along a continuum. The value of a continuous variable is not limited to the set of integers; values are limited only by the degree of accuracy of the measuring instrument. Examples include weight, blood pressure, temperature, serum cholesterol levels, and age, each of which can be measured to an arbitrary number of decimal places.

It should be clearly noted that the terms discrete and continuous refer to level of measurement and are not to be confused with the concepts of theoretical and empirical distributions. Since the observations in a population may be measurements that are either continuous or discrete, it is possible to have either continuous theoretical distributions or discrete theoretical distributions.

A discrete distribution is presented by means of a frequency diagram or simply in a frequency table.

The shape of the theoretical distribution for a continuous variable is presented by means of a smooth curve. Conceptually, this curve is a histogram whose interval widths become smaller and smaller as more observations are included. As the number of observations approaches the total number of observations in the population, the interval widths become infinitely small and form a smooth curve. The relation between the histogram and the smooth curve will be demonstrated in the next section.

The Normal Distribution

The most widely used continuous theoretical distribution is the *normal* distribution.* Its popularity is due to the fact that the empirical distributions of many naturally occurring phenomena (e.g., blood pressure, height, weight, cholesterol values) approximate the normal distribution.

Figure 3.1 presents a histogram for the cholesterol levels in normal men given in Table 2.1, Chapter 2. A smooth curve connecting the midpoints of each interval has been drawn. The shape and general features of the smooth curve reflect the characteristics of a normal curve. A frequency table based on the theoretical distribution curve can be constructed and compared to the observed frequency table. The procedure for doing this follows a discussion of the properties of a normal curve.

A normal distribution is "bell-shaped" and symmetrical about its mean μ, (i.e., 50 percent of the observations lie above the mean and 50 percent below). Approximately 68 percent of the observations lie within one standard

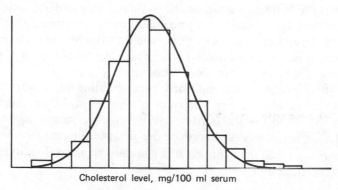

Cholesterol level, mg/100 ml serum

Figure 3.1. Frequency distribution for cholesterol levels in normal males from Table 2.1, Chapter 2.

* The word "normal" is used in several different ways. We speak of normal males when we mean free of any disease or defect relevant to our purposes; *normal values* is a commonly heard term in medicine which refers to that range of values that is routinely observed; the *normal distribution*, which properly is called the Gaussian distribution, after the mathematician Gauss, refers to the very common bell-shaped frequency distribution that is the foundation of statistical methods. The meaning of "normal" usually should be clear from the context in which it is being used.

deviation of the mean ($\mu \pm 1\sigma$) and approximately 95 percent of the observations lie within 1.96 standard deviations of the mean ($\mu \pm 1.96\sigma$). This is symbolized in Figure 3.2.

If a variable is normally distributed, then we may say with certainty that 95 percent of all the observations will fall within a range of 1.96 standard deviations of the mean μ. Likewise we may say with certainty that 99 percent of the observations will fall within 2.58 standard deviations. Stated alternatively, the probability of observing a value of a given variable in the range $\mu \pm 1.96\sigma$ is 0.95. The importance of these probabilities associated with the normal curve will become more apparent when the principles of hypothesis testing are discussed in Chapter 4.

The "location" of a normal distribution depends on the value of μ and the "spread" depends on the value of σ. See Figure 3.3.

Suppose Y is a normally distributed random variable (i.e., Y is an observation from a population whose characteristics satisfy the requirements for the theoretical normal distribution). Just as a histogram and its frequency table can be used to compute probabilities (Chapter 2), the smooth normal curve substituting for the histogram can be used to calculate the probability that Y lies in an interval between two points, a and b. The desired probability is given by the area under the appropriate frequency curve between the points a and b as shown in Figure 3.4. Table B in the Appendix is the frequency table for the standard normal distribution.

Since there are infinitely many normal distributions (recall that location and spread depend on μ and σ), observed data from any particular normal distribution must be standardized. Otherwise we would need a separate normal table for every conceivable combination of μ and σ.

Standardization is accomplished by converting the observed values to z values, or *standard scores*. The value z, also called the standard normal deviate, is the number of standard deviations an observation lies away from the mean μ. To find z, we first determine the distance between the observation Y and the mean μ and then convert to a standard score by dividing by σ (the standard deviation of Y). Hence,

$$z = \frac{\text{Distance}}{\text{Standard Deviation}} = \frac{Y - \mu}{\sigma}$$

Pictorially, the transformation from raw scores (Y) to standard scores (z) is shown in Figure 3.5. Symbolically, this is written

$$P(a \leq Y \leq b) = P(z_1 \leq z \leq z_2)$$

where

$$z_1 = \frac{a - \mu}{\sigma} \quad \text{and} \quad z_2 = \frac{b - \mu}{\sigma}$$

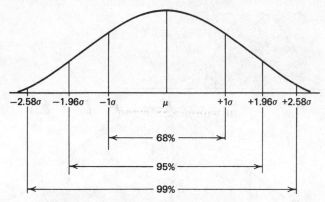

Figure 3.2. A normal distribution showing the percentage of values found in various ranges about the mean value.

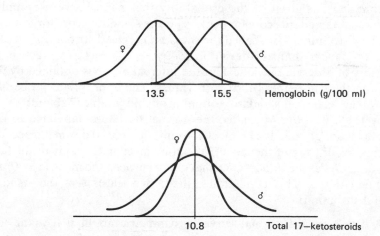

Figure 3.3. (a) Normal hemoglobin values showing the distribution around a higher mean value (μ) for males. (b) Normal 17-ketosteroid values showing a smaller spread or standard deviation (σ) for the distribution of values for females.

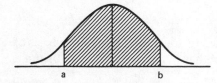

Figure 3.4. The probability that an observed Y lies between *a* and *b* is given by the shaded area.

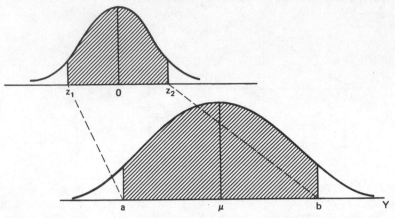

Figure 3.5.

In words, this says that the probability that a sample Y value will lie between a and b is given by the probability that z lies between z_1 and z_2 where z_1 is the standard score for a and z_2 is the standard score for b.

The distribution for z, called the *standard normal* distribution, thus has a mean of 0 and a standard deviation of 1 ($\mu = 0$, $\sigma = 1$). In practice, the probability of observing a value Y in the interval a to b is obtained by first converting the end-points $[a, b]$ to z values. The appropriate probability is then obtained from a standard normal probability table (Table B).

In Table B in the Appendix, the normal deviate z, tabulated to one decimal place, is given in the leftmost column; the second decimal place for z is given across the top of the table. The values given in the body of the table are the areas under the standard normal curve between the mean ($z = 0$) and any observation z. This tabulation utilizes two features of the standard normal distribution:

1. The standard normal curve is symmetric about 0 (i.e., the area between 0 and z_1 is the same as the area between 0 and $-z_1$).
2. The total area from 0 to $+\infty$ is 0.5. It is strongly suggested that in obtaining probabilities from Table B, the student first draw the normal curve and identify the appropriate areas of interest as was done in Figure 3.5.

Since the normal is symmetrical, areas can be tabulated in one direction only thus saving space.

In summary,

1. Statistical tables for the normal curve are constructed on the basis of the standard normal curve, which has a mean of zero ($\mu = 0$) and a standard deviation of one ($\sigma = 1$).

2. Observed data may have any mean and standard deviation, not necessarily (0, 1).
3. Raw data therefore must be reduced to standard scores (z values) in order for the normal table to be applicable. Reducing all values to z scores is analogous to finding a common denominator for a group of common fractions for purposes of comparison.
4. While raw values (Y) have been standardized, the probabilities (proportion, percent) obtained using the standard normal curve have been preserved. That is

$$P(a \leq Y \leq b) = P(z_1 \leq z \leq z_2)$$

where

$$z_1 = \frac{a - \mu}{\sigma} \quad \text{and} \quad z_2 = \frac{b - \mu}{\sigma}$$

These properties now will be illustrated by means of examples.

Example 1. Refer to the cholesterol example (Table 2.1, Chapter 2). Assuming the distribution of serum cholesterol level in normal men possesses the characteristics of the theoretical normal distribution, find the probability that a normal male will have a cholesterol value between 189.5 and 329.5. For the example we will assume that $\mu = 242.2$ and $\sigma = 45.4$.

The frequency distribution of the actual observed data is shown in Figure 3.1. Since we are assuming that the distribution of the original data (frequency histogram) satisfactorily meets the requirements of a normal distribution, then we may ignore the frequency histogram and concentrate on the smooth curve (theoretical distribution). The shaded portion of each curve in Figure 3.6 represents the area corresponding to the probability of interest.

To obtain the actual proportion of normal men with serum cholesterol levels in the interval 189.5 to 329.5, the end-points of the interval must first be converted to z values. Thus,

$$P(189.5 \leq Y \leq 329.5) = P\left(\frac{189.5 - 242.2}{45.4} \leq z \leq \frac{329.5 - 242.2}{45.4}\right)$$

$$= P(-1.16 \leq z \leq 1.92)$$

The z values and the corresponding area of interest are also shown in Figure 3.6. Note that the proportion represented by the shaded area in the top curve of Figure 3.6 is exactly the same as the proportion represented by the shaded area in the lower curve. That is, proportions (probabilities, percent) as represented by areas under the normal curves have been preserved but observations have been standardized.

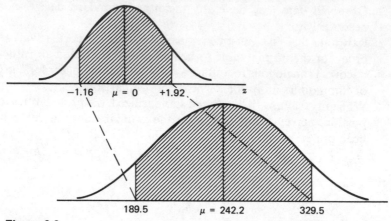

Figure 3.6.

The probability associated with the interval -1.16 to 1.92, and the interval 189.5 to 329.5, is obtained from the standard normal table. In Table B, the entry corresponding to 1.92 is 0.4726. That is, the area between 0 and $z = +1.92$ is 0.4726. Likewise, the area between 0 and $z = -1.16$ is 0.3770. (Recall that because of symmetry only positive z values are tabulated.) Figure 3.7 shows the given areas.

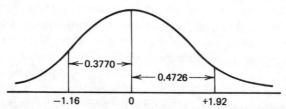

Figure 3.7. Areas corresponding to $z = +1.92$ and $z = -1.16$.

Thus, the total area between $z = -1.16$ and $z = +1.92$ is given

$$P(-1.16 \leq z \leq 1.92) = 0.3770 + 0.4726 = 0.8496.$$

The probability of obtaining a normal male with serum cholesterol value in the interval 189.5 to 329.5 is 0.8496. How does this value compare with the observed relative frequency given in Table 2.1, Chapter 2? *lower*

~.91

Example 2. Find the probability of obtaining a normal male with serum cholesterol value less than or equal to 159.5.

The probability corresponding to the shaded portion of Figure 3.8 is again obtained by converting 159.5 to a z value.

$$P(Y \le 159.5) = P\left(z \le \frac{159.5 - \overset{\mu}{242.2}}{45.4}\right)$$

$$= P(z \le -1.82)$$

The entry in Table B corresponding to the area between 0 and $z = +1.82$ is 0.4656. This value is indicated in Figure 3.8. Since the total area below $\mu = 0$ is 0.5, the area below $z = -1.82$ is obtained by subtraction

$$P(Y \le 159.5) = P(z \le -1.82) = 0.5 - 0.4656 = 0.0344.$$

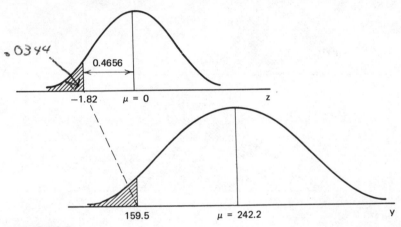

Figure 3.8.

Thus, 3.4 percent of normal men aged 40–59 have cholesterol levels less than 159.5 based on the use of the standard normal curve. That is, for a normal male aged 40–59 selected at random from this group, the probability that his cholesterol level will be less than 159.5 is 0.0344 (or 3.4 percent) based on the normal approximation. This compares favorably with the observed value of 3 percent (see Table 2.1, Chapter 2).

Example 3. Between what two cholesterol values will 90 percent of all cholesterol values for normal adult males aged 40–59 lie, such that 5 percent are above the upper limit and 5 percent are below the lower limit?

In the preceding examples we were interested in determining the probabilities corresponding to given z values. In this example the probability is given and we wish to determine the corresponding z, and hence Y, values.

Pictorially, we wish to find the z values corresponding to the shaded area of Figure 3.9.

The total area between z_1 and z_2 is 0.90; the area between 0 and z_1 or 0 and z_2 is 0.45 since the interval is centered on $\mu = 0$. To obtain a z value corresponding to an area of 0.45, we must enter the *body* of Table B, locate 0.45 (or the area nearest to 0.45), and obtain the z that corresponds to this area. Note that the area of 0.45 was used since Table B is tabulated in terms of areas from 0 to z. From Table B the z value which gives an area closest to 0.45 is 1.65. (The correct value would be 1.645 since 0.45 lies halfway between .4495 and .4505. However, for our purposes, the approximation will suffice.) Since the interval is symmetric, $z_1 = -1.65$ and $z_2 = +1.65$.

[handwritten margin note: why not .5?]

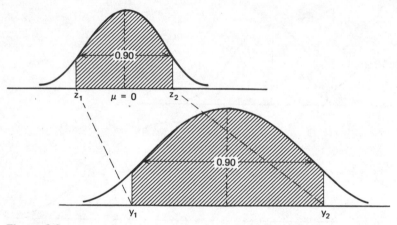

Figure 3.9.

The cholesterol values which are equivalent to the z values, $+1.65$ and -1.65, are given by

$$z = \frac{Y - \mu}{\sigma}$$

$$-1.65 = \frac{Y_1 - 242.2}{45.4} \qquad +1.65 = \frac{Y_2 - 242.2}{45.4}$$

$$Y_1 = 167.29 \qquad\qquad Y_2 = 317.11$$

Thus,

$$P(167.29 \le Y \le 317.11) = 0.90$$

Sampling Distributions
and the Central Limit Theorem

Derived distributions. In previous sections the relationship between population parameters and sample statistics has been presented. A statistic is computed from a sample and is used as an estimate of the population parameter. Not only are statistics rarely if ever equal to their corresponding population parameters, but upon repeated random sampling under identical conditions the computed statistics will vary among themselves.

Suppose, for example, that an estimate of the average length of stay in the hospital of a certain type of hospitalized patient was desired. One approach would be to sample a certain number of hospital wards for the past calendar year and use the average of the sample values as the estimate. A second sample of the same size as the first would, in all likelihood, yield a different average value, as would further samples. Thus, the series of mean values from repeated samples of the same size would have a distribution of their own, a distribution derived from the original parent distribution.

The derived distribution of sample means (sometimes called the sampling distribution of the mean) is of unique importance in statistics. In order to gain insight into the meaning and utility of sampling distributions, let us first consider how the distribution of sample means for a given sample size n could be obtained. It is important to note that this simply will be an illustrative exercise; a sampling distribution rarely is generated in practice, but a knowledge of its properties is essential for making statistical inferences discussed in the next chapter.

The steps for generating a sampling distribution of $\bar{Y}$s are as follows:

STEP 1. Construct a population by recording population values on slips of paper and placing the slips in a container.

STEP 2. Select a sample of size n from the population and compute its mean $\bar{Y}$.

STEP 3. Replace these n observations in the population.

STEP 4. Continue steps 2 and 3 until a large number of samples of size n have been drawn.

STEP 5. Record the frequency of occurrence of each of the $\bar{Y}$s. The resulting frequency distribution approximates the distribution of sample mean values, or more technically the *sampling distribution of means*, for a given n.

As you may have perceived, the process of generating a complete sampling distribution becomes an impossible task for parent populations having more than a few observations. Fortunately, one does not have to rely on such formidable techniques to obtain information about the characteristics of every sampling distribution. In general, the information of value—the mean, the standard deviation, and the shape of the distribution when graphed—may all be obtained mathematically. The mathematical derivation of these properties is beyond the scope of this text. The interested reader is referred to any theoretically oriented statistical text. We will simply state the mathematical conclusions as they are needed.

Properties of the distribution of sample means. The properties of the sampling distribution of the mean may be summarized as follows:

1. The mean $\mu_{\bar{y}}$ of the distribution of $\bar{Y}$ is equal to μ, the mean of the parent population from which the samples were drawn.
2. The standard deviation, $s_{\bar{y}}$, of the sampling distribution of $\bar{Y}$ is equal to $\sigma/\sqrt{n}$, the standard deviation of the parent population divided by the square root of the number in each sample. The standard deviation of the distribution of $\bar{Y}$, called the *standard error of the mean*, is given in terms of the standard deviation of the individual values of the parent distribution. This allows the calculation of an estimate of the standard deviation of mean values from a single sample of individual observations.
3. The shape of the sampling distribution of $\bar{Y}$ is that of the normal curve when sampling is from a parent population whose distribution is normal. The sampling distribution of $\bar{Y}$ is approximately normal when sampling is from a non-normal parent population. As the sample size increases, the approximation to normality becomes closer and closer.

This third property of the sampling distribution of $\bar{Y}$ is called the *Central Limit Theorem* and is of great importance to statistical methodology. Stated more formally, the Central Limit Theorem is as follows: Given any parent population, not necessarily normal, having mean and standard deviation σ, the sampling distribution for fixed n which is generated from this population will be approximately normally distributed with mean, μ, and standard deviation $\sigma/\sqrt{n}$. It is around this assumption of normality for the distribution of $\bar{Y}$ that much of statistical inference is built.

Summary. The concept of sampling distributions is often a difficult one to grasp, especially for students in an introductory course. The simplest application of the principles discussed in this section is in computing the probability of obtaining a sample with a specified mean value from a given population of values. Examples of this will be discussed in the methodology section. Further applications of the properties of sampling distributions will become apparent in the following section on confidence intervals and in the discussion of hypothesis testing in Chapters 4–7.

Confidence Interval for a Population Mean

An important aim in statistical methodology is to obtain information about the mean of a given population. Thus far, this information has taken the form of the sample mean $\bar{Y}$, a *point estimate* of μ. Additional information is provided by the *standard error*, or standard deviation of the distribution of $\bar{Y}$s. The standard error of the mean, $\sigma/\sqrt{n}$, denoted by $s_{\bar{Y}}$, is a measure of the "average" deviation of any $\bar{Y}$ from the true mean μ. Using the point estimate $\bar{Y}$, the standard error $\sigma/\sqrt{n}$ and the properties of the normal distribution, an *interval estimate* on μ may be constructed. This interval estimate, called a *confidence interval on* μ, gives a range of values which might reasonably contain μ. The confidence interval is constructed in such a way that the probability that the interval covers μ is as high as desired. The technique for constructing confidence intervals will be presented in the methodology section.

METHODOLOGY

Computing Probabilities Associated with Mean Values Using the Normal Distribution

The power of the Central Limit Theorem is in the fact that sample mean values from any distribution will be distributed approximately normally when the sample size is large enough. The normal distribution can then be used to compute probabilities associated with mean values.

Just as for individual observations, mean values must be converted into standard scores so that the standard normal distribution can be utilized.

Pictorially, the transformation from raw mean scores ($\bar{Y}$) to standard scores (z) is shown in Figure 3.10. The final distribution is the same as that shown in Figure 3.5.

Note that the denominator for z is now $\sigma/\sqrt{n}$ rather than σ. Recall that in the definition of z, the distance between the observation to be standardized and its mean μ is divided by the standard deviation of the population from which the observation is taken. Hence, the observation to be standardized is $\bar{Y}$ and the standard deviation of the population of $\bar{Y}$s is $\sigma/\sqrt{n}$.

Approximately 95 percent of all observations from any normal distribution fall within 1.96 standard deviations of the mean. Then, for repeated samples of size n drawn from a population, we know that 95 percent of the

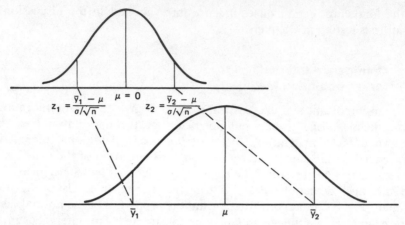

Figure 3.10.

sample means, as for any normal distribution, lie within 1.96 standard errors of the true population mean. In other words, we are 95 percent certain that any one sample mean will lie within ± 1.96 standard errors of the population mean. Similarly, the probability is 0.99 (99 percent chance) that any $\bar{Y}$ will lie within a range of ± 2.58 standard errors of the mean (see Figure 3.2). In terms of z values, 95 percent of all sample means from a given population will fall within the interval $z = -1.96$ to $z = +1.96$.

To illustrate the use of the standard normal distribution with mean values we turn once again to the serum cholesterol levels given in Table 2.1 of Chapter 2. Recall that for the population of normal men, $\mu = 242.2$ and $\sigma = 45.4$. Suppose that 25 subjects selected randomly from this population were to be used as controls in an experiment. The mean cholesterol value for this group of controls would represent a sample from the derived distribution of mean values with $\mu = 242.2$ and standard error $\sigma/\sqrt{n} = 45.4/\sqrt{25} = 9.08$.

What is the probability that the mean of the control group would be 265 mg/100 ml or greater? The shaded portion of the curve in Figure 3.11 is the area corresponding to the proportion of mean values that are greater than or equal to 265 mg/100 ml.

As before, to obtain the probability from the standard normal table corresponding to the shaded area in Figure 3.11, the mean value ($\bar{Y}$) must be converted to a standard score.

$$z = \frac{\bar{Y} - \mu}{\sigma/\sqrt{n}} = \frac{265 - 242.2}{45.4/\sqrt{25}} = 2.51$$

From Table B, the entry corresponding to $z = 2.51$ is 0.4940. The probability corresponding to the desired shaded area in Figure 3.11 is then given

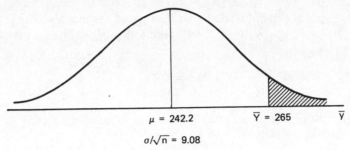

$\mu = 242.2 \qquad \bar{Y} = 265 \qquad \bar{y}$

$\sigma/\sqrt{n} = 9.08$

Figure 3.11.

by $0.5 - 0.4940 = 0.006$. Symbolically,

$$P(\bar{Y} \geq 265) = P(z \geq 2.51) = 0.006$$

Thus, there is less than a 1 percent chance that the mean serum cholesterol of a randomly selected group of normal men would be 265 mg/100 ml or greater.

Clearly, computing probabilities concerning mean values follows exactly the same procedure as outlined for individual values.

Confidence Intervals on μ

The sample mean ($\bar{Y}$) is the best available estimate of the true mean, μ. However, since sample means vary among themselves and are not necessarily identical to μ, it is sometimes desirable to indicate in some manner the precision of the point estimate, $\bar{Y}$. One way to do this is to report the standard error of the sampling distribution along with the estimate $\bar{Y}$. An alternate way of presenting information on μ is to construct an interval that depends on both the point estimate ($\bar{Y}$) and the standard error ($\sigma/\sqrt{n}$). As stated in the section on sample distributions and the Central Limit Theorem, this interval, called a confidence interval on μ, is constructed in such a way that the probability that the interval covers μ is high (usually 90 percent or greater).

Construction of confidence intervals.   As shown in the previous section, 95 percent of all $\bar{Y}$s lie within 1.96 standard errors of the true population mean μ. This probability statement in terms of standardized values may be expressed symbolically

$$P(-1.96 \leq z \leq 1.96) = 0.95$$

In words, this statement says that the probability that any z, the standard score for any $\bar{Y}$, lies in the interval -1.96 to $+1.96$ is 0.95. Substituting the term $(\bar{Y} - \mu)/(\sigma/\sqrt{n})$ for z (recall that to standardize mean values, it is necessary to divide by $\sigma/\sqrt{n}$ rather than σ) the following expression is obtained:

$$P\left(-1.96 \leq \frac{\bar{Y} - \mu}{\sigma/\sqrt{n}} \leq 1.96\right) = 0.95$$

By multiplying the inequality by $\sigma/\sqrt{n}$, subtracting $\bar{Y}$ from both sides, and finally multiplying by (-1) we obtain

$$P(\bar{Y} - 1.96\sigma/\sqrt{n} \leq \mu \leq \bar{Y} + 1.96\sigma/\sqrt{n}) = 0.95$$

The above relationship written in the form

$$CI_\mu = \bar{Y} \pm 1.96\sigma/\sqrt{n}$$

is called a 95 percent confidence interval on the true mean μ. It should be emphasized that the foregoing is merely an algebraic consequence of manipulations of standard normal deviates.

Interpretation of confidence intervals on μ.   Confidence intervals may be interpreted in two ways. The first is of a more theoretical nature and involves a consideration of what happens in the long run. If a series of samples were obtained and 95 percent confidence intervals on the true mean μ constructed for each sample, then, in the long run, the relative frequency with which the computed intervals actually cover the true population mean μ is 95 percent. In other words, if one were to construct 100 confidence intervals on μ using 100 different $\bar{Y}$s, then one would expect 95 of the intervals to actually contain μ. A second more useful interpretation is that, for the single sample obtained in practice, we are 95 percent confident that the interval $\bar{Y} \pm 1.96\sigma/\sqrt{n}$ computed from that particular sample covers the true population mean μ.

General form of confidence intervals on μ.   The general form of a confidence interval on μ is given by

$$CI_\mu = \bar{Y} \pm z_{(1-\alpha)/2}\sigma/\sqrt{n}$$

The symbols α and $(1 - \alpha)$ may best be described pictorially. The sum of the shaded areas of Figure 3.12 is called α and is of primary importance in hypothesis testing (Chapter 4). Since the total area under the curve is equal to 1, the area of the *nonshaded* portion of Figure 3.12 is $(1 - \alpha)$. The value $(1 - \alpha)$ is the *confidence coefficient*, or level of confidence associated with a

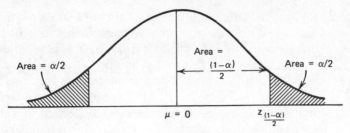

Figure 3.12.

confidence interval. The value $z_{(1-\alpha)/2}$ given in the general formula for confidence intervals is the z score obtained from Table B which corresponds to the area $(1-\alpha)/2$. For a 95 percent confidence interval, the confidence coefficient, $(1-\alpha)$, is 0.95 and the appropriate z value corresponding to an area of $(1-\alpha)/2$ is 1.96. Verify that for confidence coefficients of 0.99 and 0.98 the $z_{(1-\alpha)/2}$ values are 2.58 and 2.33, respectively.

Example of a confidence interval on μ. In a study to determine the effect of stress on freshmen medical students, the systolic blood pressures of 50 freshmen students were taken immediately following a mid-term exam on gross anatomy. The mean systolic blood pressure for the 50 students was found to be 150 mmHg. Assume the standard deviation (σ) of systolic blood pressures is equal to 20 mmHg.

A 95 percent confidence interval on the *true mean* systolic blood pressure reading for all the freshmen following the gross anatomy exam is given by

$$CI_\mu = \bar{Y} \pm z_{.475}\sigma/\sqrt{n}$$

$$= 150 \pm 1.96\frac{20}{\sqrt{50}}$$

$$= [144.5, 155.5]$$

We are 95 percent confident that the interval 144.5 to 155.5 mmHg covers the true mean systolic blood pressure for all freshmen taking the exam.

The Student-t Distribution

The confidence intervals discussed in the previous section involved the population parameter σ. Rarely in practice, however, is the value of σ actually known. The most logical solution to the problem of an unknown σ is to use the estimate s calculated from the sample. The obvious problem with the substitution of s for σ is that in addition to the variability of $\bar{Y}$, s now also varies with each sample. This additional variability in s is taken into account by use of the *student-t distribution* (in common usage simply the t distribution) in place of the z distribution described earlier. The t distribution, while similar to the z in most respects, requires an additional calculation called the *degrees of freedom*, symbolized df. The degrees of freedom are equal to $n - 1$ where n is the number of observations in the sample. For very large n, the t and z distributions become indistinguishable. (Notice that in Table C in the Appendix the t values in the row labeled $n = \infty$ are the familiar z scores.) In the remainder of this text the following criteria will be used to distinguish between the use of z and t: Use z when the population standard deviation σ is known; use t when σ is unknown.

Steps for obtaining values from the t table.   As stated earlier, the t distribution requires the calculation of a value called degrees of freedom (df) where $df = n - 1$. Since the shape of the t distribution depends upon the sample size n, there is a different t distribution for each n. A complete listing of each t value for all possible sample sizes therefore would be quite impractical. In most statistical work, a condensed t-table consisting of t values corresponding to the upper tail areas of 0.10, 0.05, 0.025, 0.01, and 0.005 is sufficient. In Table C the degrees of freedom are listed in the leftmost column and the upper tail areas (α) are given across the top row of the table. Consider Figure 3.13. For $n = 5$ and $\alpha = 0.05$, the appropriate t value from Table C is 2.132. This value was found by locating the intersection of the row $df = 4$ and the column $\alpha = 0.05$. Note that a z value is given in terms of the area between 0 and z; a t value is given in terms of the area to the *right* of the point t.

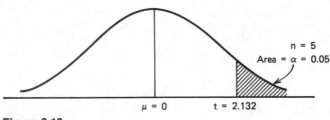

$\mu = 0$ $t = 2.132$

n = 5
Area = α = 0.05

Figure 3.13.

Confidence Intervals Using the t Distribution

The rationale and methodology for confidence intervals are the same for both the t and z distributions. For a CI using t, the estimate s is substituted for σ and the appropriate t value is obtained from Table C. The general form for a confidence interval on μ using t is

$$CI_{\mu} = \bar{Y} \pm t_{\alpha/2}s/\sqrt{n}$$

The appropriate t value for a 95 percent CI on μ for $n = 5$ would be 2.776. Note in Figure 3.14 that for a 95 percent level of confidence, $\alpha = 0.05$ and $\alpha/2 = 0.025$. When placing confidence intervals on μ using t an area equal to $\alpha/2$ is placed in each tail of the curve as was the case when using z.

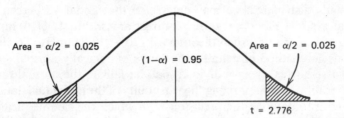

Figure 3.14.

Summary. The steps for placing confidence intervals on μ are as follows:

STEP 1. Decide which confidence level is to be used (usually 95 percent or 99 percent)

STEP 2. Determine $\bar{Y}$ and s (if σ unknown) from the sample and calculate the standard error of the mean

$$s_{\bar{y}} = s/\sqrt{n} \text{ (or } \sigma/\sqrt{n} \text{ if } \sigma \text{ is known)}$$

STEP 3. For σ known,

$$CI_{\mu} = \bar{Y} \pm z_{(1-\alpha)/2}\sigma/\sqrt{n}$$

For σ unknown,

$$CI_{\mu} = \bar{Y} \pm t_{\alpha/2}s/\sqrt{n}$$

STEP 4. Interpretation—For a 95 percent CI on μ, we can assert that we are 95 percent confident that the given interval covers the true population mean μ.

The Difference Between the Standard Deviation and the Standard Error of the Mean

Often in the medical literature there is confusion as to whether the standard deviation of the observations (s) or the standard error of the mean ($s/\sqrt{n}$) is being reported. When, for example, the statement 120 ± 10 mmHg is given to summarize the systolic blood pressures of a group of 25 individuals, one is not sure exactly which measure of the spread of the data is meant. The distinction between the two is critical because the standard deviation (s) and the standard error ($s/\sqrt{n}$) are used for quite different purposes.

When it is of interest to describe the variability of a set of *individual* observations, the standard deviation, s, is appropriate. For example, suppose 25 systolic blood pressure measurements are summarized as 120 ± 10 mmHg, where $\bar{Y} = 120$ is the mean of the 25 observations and $s = 10$ is the standard deviation of the set of 25 observations. Assuming that systolic blood pressures are normally distributed, we may determine that about 95 percent of the individual systolic blood pressure readings are within 1.96(10) mmHg of the mean of 120, that is, are in the interval 120 ± 19.6 mmHg. To clinicians this is a useful method of defining the range of clinically normal values so they can assess whether an individual patient falls in the normal range.

In addition to summarizing the variability of the individual data values, we may wish to describe the precision with which the sample mean $\bar{Y}$ has been determined. As was discussed in the preceeding section, the dispersion of the population of sample means ($\bar{Y}$s) about the true population mean μ is given by the standard error of the mean, $\sigma/\sqrt{n}$. Thus, to describe the precision with which an individual sample mean $\bar{Y}$ estimates the true population mean μ, we may use the form $\bar{Y} \pm s/\sqrt{n}$, where $s/\sqrt{n}$ is an estimate of the true standard error, $\sigma/\sqrt{n}$. To the researcher this formulation is useful in assessing experimental error and making judgments about the validity of the measurement process.

Often, authors of the medical literature do not make it clear when giving a statement such as 120 ± 10 mmHg whether they are giving $\bar{Y} \pm s$ or $\bar{Y} \pm s/\sqrt{n}$. Since standard error of the mean is smaller than the standard deviation of individual observations, incorrect conclusions about the variability of the author's data may be drawn if s and $s/\sqrt{n}$ are incorrectly interchanged.

In summary, if we wish to describe variability of the set of individual observations, $\bar{Y} \pm s$ is appropriate. The standard error of the mean is used in $\bar{Y} \pm s/\sqrt{n}$ to describe the precision of $\bar{Y}$ as an estimate of the true population mean μ.

PROBLEMS

1. For the cholesterol example, find: ($\mu = 242.2$, $\sigma = 45.4$)
   a. $P(219.5 \leq Y \leq 259.5)$
   b. $P(139.5 \leq Y \leq 219.5)$
   c. $P(159.5 \leq Y \leq 179.5)$
   d. $P(Y \geq 242.2)$
   e. $P(Y \leq 219.5)$

2. Using the standard normal table, what is the area within ± 1.96 standard deviations of the mean? Within ± 2.58? How does this relate to the second characteristic of the normal distribution as listed in the section on the normal distribution?

3. The plasma potassium levels (in milliequivalents per liter) were obtained for 30 adult males with a certain disease. The distribution of the 30 readings is given below.

Plasma Potassium Level	Frequency	Relative Frequency	Theoretical Frequency
2.35–2.55	1	.033	
2.55–2.75	3	.100	
2.75–2.95	2	.067	
2.95–3.15	4	.133	$\bar{Y} = 3.35$
3.15–3.35	5	.167	$s = 0.50$
3.35–3.55	6	.200	
3.55–3.75	3	.100	
3.75–3.95	3	.100	
3.95–4.15	2	.067	
4.15–4.35	1	.033	

It is known that plasma potassium levels for the entire population of values follow the normal distribution, that is, if it were possible to obtain *all* plasma potassium measurements for the entire population of adult males with the disease, then their distribution would be representable by the normal curve.

a. Using the normal distribution as the theoretical model for the empirical frequency table given above, complete the theoretical frequency column. (Use $\bar{Y}$ and s as estimates of μ and σ). Note that the sum of the theoretical frequency column will not be exactly 1.00 since the areas below 2.35 and above 4.35 were ignored. Why are these values referred to as "theoretical" frequencies?

b. If the sample size is sufficiently large and there is a large difference between the observed (empirical) and the theoretical values, what might this tell us about the assumption of normality?

4. Suppose a clinically accepted value for mean systolic blood pressure in males aged 20–24 is 120 mmHg and the standard deviation is 20 mmHg.

a. If a 22-year-old male is selected at random from the population, what is the probability that his systolic blood pressure is equal to or less than 150 mmHg? Equal to or less than 110?

b. The systolic blood pressure of a 20-year-old male selected at random from the population was 160. How many standard deviations above the mean is this value? (Hint: z values are in units of standard deviations.)

    c. Between what two blood pressure readings will 95 percent of all systolic blood pressure readings for 20–24-year-old males lie? Between what two values will 90 percent of the readings lie?

    d. What proportion of readings lie within the range 100–140 mmHg?

    e. What proportion will lie outside the range 60–180 mmHg?

5. Refer to problem 4. As part of a screening project, the systolic blood pressures of 100 freshmen medical students were taken. The mean systolic blood pressure of this sample was 126 mmHg.

    a. For $n = 100$, what is the probability of obtaining a sample mean value equal to or greater than 126 mmHg from a population with true mean 120 mmHg? (Hint: We are no longer looking at the distribution of the actual systolic blood pressure readings but rather at the distribution of mean ($\bar{Y}$) systolic blood pressure readings from samples of size 100 drawn from the above population. According to the Central Limit Theorem, what are the mean and standard deviation of this sampling distribution of $\bar{Y}$s?)

    b. How many standard deviations above the true mean is the observed mean value?

    c. Between what two mean values would we expect 95 percent of all sample means (of size $n = 100$) to lie?

    d. Based on the above information and computations, what conclusions might be drawn concerning the systolic blood pressure of medical students?

    e. Above what value would 10 percent of the sample means ($n = 100$) lie?

    f. What is the probability of selecting a sample with mean equal to or less than 115 mmHg from a population with true mean systolic blood pressure equal to 120 mmHg?

6. a. In a study to determine the effect of bumetanide on urinary calcium excretion, nine randomly selected males were each given an oral dose of 0.5 mg of the drug. Urine was collected hourly for 6 hours. The mean excretion rate for this sample of nine males was found to be 7.5 mg/hour with a standard deviation of 6.0 mg/hour. Place a 95 percent confidence interval on the true mean excretion rate for all males on the drug.

    b. Urine collections were also made for 16 randomly selected males not taking the medication. The mean of this sample was 6.5 mg/hour and the standard deviation was 2.0 mg/hour. Place a 95 percent confidence interval on the true mean excretion rate for males not on the medication.

7. Suppose a clinically accepted value for serum cholesterol level in normal healthy males in the general population is 240mg/100 ml serum and the

standard deviation is 40 mg/100 ml. Cholesterol measurements were made on each of 100 males who have suffered a definite coronary event during a specified two-year period. The mean serum cholesterol level for this group was 260 mg/100 ml.

a. Place a 95 percent confidence interval on the true mean serum cholesterol level of the diseased group.

b. Based on this information, can you say that mean serum cholesterol level for the disease group is different than that for the nondisease group?

8. Refer to problem 5. Place a 99 percent confidence interval on the true mean systolic blood pressure for all freshmen medical students.

CHAPTER 4

TESTS OF HYPOTHESES ON POPULATION MEANS

OVERVIEW

Estimation of a population mean μ by using the information in the sample mean $\bar{Y}$ has been discussed in the previous chapters. It was shown that the $\bar{Y}$s are rarely equal to μ because of sampling error. The sampling distribution of $\bar{Y}$ was used to make better estimates of μ than could be obtained by using $\bar{Y}$ alone and acting as if it were the true mean of the population.

There are two general methods used to make a "good guess" as to the true value of μ. The first, discussed in Chapter 3, involves determining a range in which μ will most likely be found. This range is commonly referred to as a Confidence Interval (CI) on μ. The second method will be concerned with making a guess as to the value of μ and then "testing" to see if such a guess is compatible with the observed data. This method, called hypothesis testing, will be described in this chapter.

OBJECTIVES

- To test a hypothesis about the mean of a population using data from a single sample
- To place confidence intervals on the mean of a population using data from a single sample
- To test the significance of the difference between two means using the z-test
- To test the significance of the difference between two means using the paired t-test
- To test the significance of the difference between two means using the two-sample (pooled) t-test

- To define Type I and Type II errors
- To define power of a statistical test
- To place confidence intervals on differences between two means
- To choose the appropriate test of significance given an experimental situation

FOUNDATIONS

Introduction

Statistical hypothesis testing usually takes the following form:

1. A single population is hypothesized to have a certain mean; or, two or more populations are hypothesized to have equal mean values.
2. If two or more populations are involved in the hypothesis, statistical tests require that each population has the same standard deviation. In most instances this assumption is justifiable.
3. Differences among sample means, or the difference between a single mean and its hypothesized value, are compared to the standard error to determine, in terms of probability, how "large" the observed difference is.
4. The null hypothesis of no difference among population mean values is rejected or not rejected on the basis of the probability associated with the observed sample mean values.

Several features of this approach to experimentation are worth noting. First, the null hypothesis is merely a mechanism that allows probability levels to be calculated. That is, we temporarily act as if the true sampling distribution were known. The experimenter does not necessarily either want or expect the null hypothesis to be true. It is just a means to an end. Second, differences among means are compared to the standard error calculated (or known) from the data. This places the error associated with experimental technique and the natural variation among experimental units in a central and dominating position in statistics. Third, decisions are made in terms of probability. No absolute conclusions can ever be reached and there is always the risk of error. These principles will be demonstrated in the following sections.

One-sample Hypothesis Tests
Using z

Consider the following hypothetical example.

A research worker wishes to determine the effect of a certain experimental diet on serum cholesterol level in normal men aged 40–59. From experience he knows that the average serum cholesterol level for this population is approximately 242 mg/100 ml serum ($\mu = 242$) with a standard deviation of 45 ($\sigma = 45$) (see Table 2.1, Chapter 2). He wishes to determine whether the experimental diet has changed the serum cholesterol level, that is, is the mean cholesterol level under the diet the same as that for the normal group?

A possible approach to answering this question is to give a sample of men from this age range the diet and then compute $\bar{Y}$ for the sample to see if it is equal to the mean of the normal group ($\mu = 242$). However, from Chapter 3, we know that due to sampling variation, the $\bar{Y}$s vary among themselves and around the true mean μ. Fortunately, because of the Central Limit Theorem, we know the ranges in which one can reasonably expect $\bar{Y}$ to fall, for example, 95 percent of the $\bar{Y}$s fall within 1.96 standard errors of the true mean and 99 percent of the $\bar{Y}$s fall within 2.58 standard errors of the true mean.

Using the above information, we may now structure the test of hypothesis. Conceptually, a whole population of serum cholesterol levels for men on the special diet exists although in practice only the subjects in the sample are on the diet. The mean of this experimental population is unknown and its determination is the object of the study.

To provide the framework for the statistical test, we must assume that the diet has no effect on the standard deviation of serum cholesterol levels; thus the σ for the diet population is taken to be 45. Likewise, based on his knowledge of the distribution of "normal" men, the researcher assumes temporarily that the mean of the diet population is 242. This assumption stated formally is called the null hypothesis (denoted by H_0). Using the null hypothesis, H_0, the probability associated with a given sample mean, $\bar{Y}$, can be computed as was done in Chapter 3. The ultimate aim of a test of hypothesis is to be able to reach a decision on H_0 based on information in the sample. H_0 is rejected if the $\bar{Y}$ from the sample has associated with it a small probability of occurring; then we may say that diet does affect serum cholesterol level in normal men aged 40–59. In effect, we are saying that the likelihood of getting such a mean value by random chance is too small for us to believe that the null hypothesis is true. If, on the other hand, H_0 is not rejected because the probability associated with the observed $\bar{Y}$ is large and thus such samples are common, then it must be concluded that there is not enough evidence to show that diet affects cholesterol level. The probability levels commonly used to test null hypotheses are 5 percent and 1 percent.

Suppose the researcher calculates a mean cholesterol level of 200 for an experimental group of 81 subjects. Notice that there is quite a difference between $\mu = 242$ and $\bar{Y} = 200$. There are two possible explanations: (1) The true mean of the diet group is 242 and a very unusual sample has been drawn (recall the principles of sampling error); or (2) the *true* mean of the population of dieters is not equal to 242. To discern which of these two explanations is the more likely, we estimate the probability of obtaining a sample mean ($\bar{Y}$) equal to or smaller than 200 when the null hypothesis is true, that is, when μ equals 242.

First, we will convert $\bar{Y}$ to a "standard score." Recall

$$z = \frac{\text{distance between the observed mean and } \mu}{\text{standard error}}, \text{ or}$$

$$z = \frac{\bar{Y} - \mu}{\sigma/\sqrt{n}}$$

$$z = \frac{200 - 242}{45/\sqrt{81}} = \frac{-42}{5} = -8.4$$

The z value we calculate will give a measure of the "distance" between $\bar{Y}$ and μ in units of standard error. Recall also the probabilities depicted in Figure 4.1. Where does the z value of -8.4 fall on the scale in Figure 4.1? We know that 95 percent of all $\bar{Y}$s will fall between -1.96 and $+1.96$ due to random chance (sampling error) alone. Since the z value of -8.4 falls outside this range, the probability of observing a $\bar{Y}$ this far from its true population mean is less than 5 percent, and thus it is likely that the population mean is not 242. It is still possible that the true mean is 242 and in reality we have drawn an unusual sample. However, the probability is less than 5 percent that this is the case. The researcher cannot know that the true mean μ is not 242, since he has only a sample from the population on which to base his decision. But, the conclusion that μ is not 242 is "more likely" than that such a very unusual sample was drawn. Thus, he "rejects" the null hypothesis that $\mu = 242$ and concludes that diet does affect the

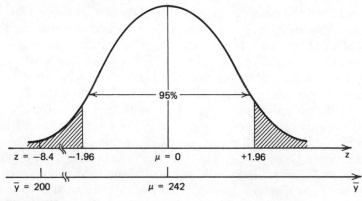

Figure 4.1.

serum cholesterol level of "normal" men aged 40–59. When the null hypothesis is rejected, the investigator has a 5 percent chance of having made an incorrect statistical decision. That is, if the null hypothesis is true, there is a 5 percent chance of erroneously rejecting it. A special name is given to this 5 percent chance of making an incorrect decision (i.e., rejecting H_0 when in reality it is true). This term is called *level of significance* and is denoted by the symbol α (here, $\alpha = 0.05$). Although the value chosen for α is arbitrary, the most commonly used values are $\alpha = 0.05$ and $\alpha = 0.01$.

It should be pointed out that in correct statistical terminology a hypothesis is either "rejected" or "not rejected" rather than "accepted." To accept the null hypothesis would imply erroneously that we are able to state definitely that the true mean of the population is 242. To be correct, we may only say that we do not have evidence to show that the true mean is not equal to the hypothesized value. A more detailed discussion of statistical decisions and their outcomes is given later in this chapter.

One-sample Hypothesis Tests Using t

In the above section, the standard deviation σ was known. In actuality, σ is rarely known and s must be calculated from the sample and used in place of σ. When the population standard deviation is not known, the t distribution is employed. The theory underlying tests of hypothesis is identical for z and t. In practice, rather than defining the distance between $\bar{Y}$ and μ in units of z, units of t are used instead.

$$t = \frac{\bar{Y} - \mu}{s/\sqrt{n}}$$

In addition, the probability "cut-off" points for determining whether the distance between $\bar{Y}$ and μ is significant also must be defined in terms of t. For a level of significance of 0.05, we determine the t value corresponding to the shaded area of Figure 4.2.

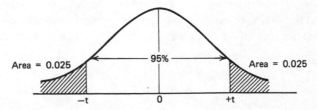

Figure 4.2. The t distribution.

Recall that t values in the table depend on the size of the sample ($df = n - 1$). To obtain the t value, find the row in Table C in the Appendix corresponding to the appropriate df and select the value for t under the column $t_{.025}$ ($\alpha = 0.05$). Verify that for $n = 10$ the value is ± 2.262. Note that the t value is larger than the corresponding z value (± 1.96). This is because we have less information (we do not know σ) and thus have a larger interval of uncertainty.

Hypothesis Test on Difference Between Two Means Using z

The hypothesis tests described in the previous sections are called one-sample tests since data from only one sample were involved. Suppose, in contrast to the example in the section on one-sample tests using z, the investigator has only a sample from the normal (nondiet) group together with a sample from the experimental (diet) group. The question to be answered is, "Is the mean serum cholesterol level the same for the diet and the nondiet groups?" Stated more formally, the null hypothesis is $\mu_1 = \mu_2$, or $\mu_1 - \mu_2 = 0$. A random sample from both the diet and nondiet groups is taken and the sample mean $\bar{Y}_1$ and $\bar{Y}_2$ for both groups is computed. The magnitude of the difference, $\bar{Y}_1 - \bar{Y}_2$, may serve as an indicator of whether there is a true difference in the population means μ_1 and μ_2 or whether the observed difference in sample means may be attributable to chance (sampling variation). As before, a "large" difference between the hypothesized population value ($\mu_1 - \mu_2 = 0$) and the observed sample estimate ($\bar{Y}_1 - \bar{Y}_2$) indicates that it is likely that the hypothesized population value is not correct; that is, it is likely that there is a difference in population means. A "small" difference means that it is likely that random chance alone accounts for the discrepancy between the observed sample value and the hypothesized population value. In this latter case, the data do not provide sufficient evidence to refute the claim of equality of population means.

The reasoning invoked in the one-sample tests of significance may be extended to the two-sample case due to the fact that if $\bar{Y}_1$ and $\bar{Y}_2$ are both distributed normally, then the difference $\bar{Y}_1 - \bar{Y}_2$ also is distributed normally. Furthermore, if the distribution of $\bar{Y}_1$ has a mean μ_1 and standard error $\sigma_1/\sqrt{n_1}$ and the distribution of $\bar{Y}_2$ has a mean μ_2 and standard error $\sigma_2/\sqrt{n_2}$, then the distribution of $\bar{Y}_1 - \bar{Y}_2$ has a mean $\mu_1 - \mu_2$ and standard error

$$\sqrt{\frac{\sigma_1^2}{n_1} + \frac{\sigma_2^2}{n_2}}$$

The standard normal deviate can be computed as before:

$$z = \frac{(\bar{Y}_1 - \bar{Y}_2) - (\mu_1 - \mu_2)}{\sqrt{\dfrac{\sigma_1^2}{n_1} + \dfrac{\sigma_2^2}{n_2}}}$$

For the special case that $\sigma_1^2 = \sigma_2^2 = \sigma^2$ we have

$$z = \frac{(\bar{Y}_1 - \bar{Y}_2) - (\mu_1 - \mu_2)}{\sqrt{\dfrac{\sigma^2}{n_1} + \dfrac{\sigma^2}{n_2}}}$$

$$= \frac{(\bar{Y}_1 - \bar{Y}_2) - (\mu_1 - \mu_2)}{\sqrt{\sigma^2 \left(\dfrac{1}{n_1} + \dfrac{1}{n_2} \right)}}$$

$$= \frac{(\bar{Y}_1 - \bar{Y}_2) - (\mu_1 - \mu_2)}{\sigma \sqrt{\dfrac{1}{n_1} + \dfrac{1}{n_2}}}$$

For the null hypothesis that $\mu_1 = \mu_2$ assuming $\sigma_1 = \sigma_2 = \sigma$, z takes the simplified form

$$z = \frac{\bar{Y}_1 - \bar{Y}_2}{\sigma \sqrt{\dfrac{1}{n_1} + \dfrac{1}{n_2}}}$$

To return to the comparison of the diet group with the nondiet group, if the calculated value for z falls outside the ± 1.96 range (for $\alpha = 0.05$), then it is likely that the difference is not the result of sample variation. That is, it is unlikely to have occurred due to random chance alone. The difference is then said to be significant, and the null hypothesis of equal population means is rejected. On the other hand, if the calculated z falls within the ± 1.96 range, we must conclude that we do not have enough evidence to reject the hypothesis of equal population means (i.e., H_0 not rejected) at the $\alpha = 0.05$ level of significance.

Confidence Interval on the Difference Between Two Population Means ($\mu_1 - \mu_2$) Using z

Suppose instead of performing a hypothesis test on the significance of the difference between two population means, we wish to estimate the difference $(\mu_1 - \mu_2)$ and attach a level of confidence to this estimate. As in the one-sample case, this procedure is called placing a confidence interval (CI) on the difference $(\mu_1 - \mu_2)$. The best available estimate for the difference $(\mu_1 - \mu_2)$ is $(\bar{Y}_1 - \bar{Y}_2)$. Using the same general form for CI employed in the one-sample case, a CI on the difference may be defined as follows:

$$\text{CI}_{(\mu_1 - \mu_2)} = (\bar{Y}_1 - \bar{Y}_2) \pm z_{(1-\alpha)/2}\sigma \sqrt{\frac{1}{n_1} + \frac{1}{n_2}}$$

It will be recognized that this still corresponds to the general form of a CI given by

("mean") $\pm z_{(1-\alpha)/2}$ (standard error of "mean")

Hypothesis Tests on the Difference Between Two Population Means $(\mu_1 - \mu_2)$ Using t

Recall that the t distribution is employed when the true standard deviation σ of the population is unknown.

Two types of t-tests for testing significance of differences between means will be presented: the *pooled* t-test and the *paired* t-test. The distinction between these two lies in the method by which the samples are drawn.

The pooled t-test. The characteristic feature of the pooled t-test is that the individual samples represent *independent* random samples from their respective populations. For example, in testing the effects of a new drug, an investigator may assign individuals at random to the treatment group and to the control group. Observations made on individuals in the treatment group are independent of those made on individuals in the control group. Under the null hypothesis, there would be no difference in population means for the two groups. For the pooled case, the number of individuals in the two samples need not be the same.

The pooled t-test is analogous to the z-test described previously for $\mu_1 = \mu_2$. The value for t is given by

$$t = \frac{(\bar{Y}_1 - \bar{Y}_2)}{s_p \sqrt{\frac{1}{n_1} + \frac{1}{n_2}}}$$

where s_p is called the pooled standard deviation.

To determine the standard error of the distribution of differences it is first necessary to find estimates of the standard deviations of the two populations. (Recall that we assume σ to be the same for both populations.) Using the sample information, we may obtain the estimates of the variances s_1^2 and s_2^2 and pool this information to obtain an estimate of the common variance. Without proof, we will state that an estimate of the pooled standard deviation is given by

$$s_p = \sqrt{\frac{(n_1 - 1)s_1^2 + (n_2 - 1)s_2^2}{n_1 + n_2 - 2}}$$

where s_1^2, s_2^2, n_1, and n_2 are the variance and sample size of each sample, respectively. The standard error of the distribution of differences is thus

$$s_p \sqrt{\frac{1}{n_1} + \frac{1}{n_2}}$$

To obtain a value for $t_{\alpha/2}$ from Table C in the Appendix, the degrees of freedom must be specified. For the pooled case, $df = n_1 + n_2 - 2$.

The paired t-test.   For the paired case, as the name implies, pairs are randomly selected from a single population. Each member of a pair is randomly assigned to one of the two treatments. The null hypothesis is that the mean difference among pairs is zero. The list of sample differences among pairs in effect acts as a single sample.

A classic example of the paired concept is the use of identical twins; one twin receives treatment A and the other twin receives treatment B. Many factors other than the treatments under study may influence the outcome of an experiment (e.g., environment, heredity). If pairings are assigned on the basis of these extraneous factors, then the effects of these factors are partially eliminated. For example, if one is interested in testing a particular teaching method and if pairings of identical twins are used, then the subjects are as nearly identical as possible (e.g., same heredity, same family life, same previous educational experiences). The observed difference therefore for the most part can be attributed to the teaching method alone.

Another widely used example of pairing observations is the "before" and "after" measurements on the same individual. This is an example of *self-pairing*, in which each individual serves as his own control.

The obvious advantage inherent in the paired situation is the partial elimination of extraneous sources of variation. This reduction in biological and other sources of variability results in a more precise comparison of sample means.

Consider the following:

Sample 1	Sample 2	Difference
a_1 ——— a_2		$a_1 - a_2 = d_1$
b_1 ——— b_2		$b_1 - b_2 = d_2$
c_1 ——— c_2		$c_1 - c_2 = d_3$
$\vdots$		$\vdots$

$$\bar{d} = \frac{\sum d}{n}$$

Since the observations are *paired* (i.e., they are hopefully nearly identical in all areas except the one under study), the differences between each pair of observations may be taken. In contrast, in the pooled case the two samples are independently selected therefore observation a_1 has no relation to observation a_2 and the taking of the difference $a_1 - a_2$ would be meaningless. In the pooled case the null hypothesis is $\mu_1 = \mu_2$, or identically $\mu_1 - \mu_2 = 0$. We tested this hypothesis by finding the $\bar{Y}$s for both samples and calculating the difference $\bar{Y}_1 - \bar{Y}_2$. The paired case is somewhat different. The null hypothesis is whether $\mu_1 - \mu_2 = D$. Usually, the hypothesis is $H_0 : D = 0$, that is, the true mean of the population of differences is equal to zero. The best estimate of the true mean (D) of the population of differences is $\bar{d}$, the mean of the column of sample differences. As in the one-sample t-test, the t value is calculated as

$$t = \frac{\bar{d} - D}{s_d / \sqrt{n}}$$

where

$$s_d = \sqrt{\frac{\sum d_i^2 - \frac{(\sum d_i)^2}{n}}{n - 1}}$$

As you may have noticed, the paired t-test is in reality a one-sample t-test performed on the column of differences.

For the paired t-test the degrees of freedom is, as for the one-sample case, equal to $n - 1$ where n is the number of pairs of observations.

When the null hypothesis is that $D = 0$ we have

$$t = \frac{\bar{d}}{s_d / \sqrt{n}}$$

which can be compared to the value in Table C in the Appendix.

Confidence Interval
on the Difference Between
Two Population Means Using t

Pooled t. As for confidence intervals using the two sample z statistic, the CI for the difference $(\mu_1 - \mu_2)$ using t is defined as

$$\text{CI}_{(\mu_1 - \mu_2)} = (\bar{Y}_1 - \bar{Y}_2) \pm t_{\alpha/2} s_p \sqrt{\frac{1}{n_1} + \frac{1}{n_2}}$$

where $t_{\alpha/2}$ has $n_1 + n_2 - 2$ degrees of freedom.

Paired t. A confidence interval for this case is again analogous to the one-sample t case. Thus,

$$\text{CI}_{(\mu_1 - \mu_2)} = \bar{d} \pm t_{\alpha/2} s_d/\sqrt{n}$$

where $t_{\alpha/2}$ has $n - 1$ degrees of freedom.

Summary: One-sample Case

1. σ known,
   a. $H_0 : \mu = \mu_0$
   b. Test statistic

$$z = \frac{\bar{Y} - \mu_0}{\sigma/\sqrt{n}}$$

   c. Rejection region
      For $\alpha = 0.05 : \pm 1.96$
      For $\alpha = 0.01 : \pm 2.58$

   d. $\text{CI}_\mu = \bar{Y} \pm z_{(1-\alpha)/2} \sigma/\sqrt{n}$

2. σ unknown
   a. $H_0 : \mu = \mu_0$
   b. Test statistic

$$t = \frac{\bar{Y} - \mu_0}{s/\sqrt{n}}$$

   c. Rejection region

$$t_{\alpha/2}, df = n - 1$$

   d. $\text{CI}_\mu = \bar{Y} \pm t_{\alpha/2} s/\sqrt{n}, df = n - 1$

Summary: Two-sample Case

1. σ known,
   a. $H_0: \mu_1 = \mu_2$, or $\mu_1 - \mu_2 = 0$
   b. Test statistic

$$z = \frac{(\bar{Y}_1 - \bar{Y}_2)}{\sigma \sqrt{\dfrac{1}{n_1} + \dfrac{1}{n_2}}}$$

   c. Rejection region
   For $\alpha = 0.05: \pm 1.96$
   For $\alpha = 0.01: \pm 2.58$

   d. $\text{CI}_{(\mu_1 - \mu_2)} = (\bar{Y}_1 - \bar{Y}_2) \pm z_{(1-\alpha)/2}\sigma \sqrt{\dfrac{1}{n_1} + \dfrac{1}{n_2}}$

2. σ unknown

Pooled t.

   a. $H_0: \mu_1 = \mu_2$, or $\mu_1 - \mu_2 = 0$
   b. Test statistic

$$t = \frac{\bar{Y}_1 - \bar{Y}_2}{s_p \sqrt{\dfrac{1}{n_1} + \dfrac{1}{n_2}}}$$

   where

$$s_p = \sqrt{\frac{(n_1 - 1)s_1^2 + (n_2 - 1)s_2^2}{n_1 + n_2 - 2}}$$

   c. Rejection region

   $t_{\alpha/2}$ and $df = n_1 + n_2 - 2$

   d. $\text{CI}_{(\mu_1 - \mu_2)} = (\bar{Y}_1 - \bar{Y}_2) \pm t_{\alpha/2}s_p \sqrt{\dfrac{1}{n_1} + \dfrac{1}{n_2}}$

   $df = n_1 + n_2 - 2$

Paired t.

   a. $H_0: D = 0$
   b. Test statistic

$$t = \frac{\bar{d}}{s_d/\sqrt{n}}$$

where

$$s_d = \sqrt{\dfrac{\sum d_i^2 - \dfrac{(\sum d_i)^2}{n}}{n - 1}}$$

c. Rejection region

$t_{\alpha/2}$ and $df = n - 1$

d. $\text{CI}_{(\mu_1 - \mu_2)} = \bar{d} \pm t_{\alpha/2} s_d / \sqrt{n}$, $df = n - 1$

One-sided Test of Hypothesis

In the statistical test of hypothesis described thus far we have been concerned with testing the null hypothesis $H_0 : \mu = \mu_0$ against the alternative hypothesis $\mu \neq \mu_0$, or in the two population case, $H_0 : \mu_1 - \mu_2 = 0$ against $\mu_1 - \mu_2 \neq 0$. This type of test is called a two-sided test since we are interested in detecting significant differences in either the positive or the negative direction. To determine the critical range (rejection region) for a two-sided test, we find the critical points such that an area of $\alpha/2$ lies in the upper tail of the distribution (either z or t) and $\alpha/2$ lies in the lower tail. See Figure 4.3.

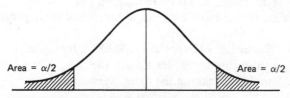

Area = $\alpha/2$ Area = $\alpha/2$

Figure 4.3. Two-tailed rejection regions.

Often the researcher is interested not only in whether $\mu \neq \mu_0$ but also in the direction of the difference. For example, he may wish to show that the mean weight of a diet group is less than the mean weight of a control group. The hypothesis to be tested is $H_0 : \mu_D = \mu_C$ against the alternative $H_a : \mu_D < \mu_C$, or rewritten, $H_0 : \mu_D - \mu_C = 0$ against $H_a : \mu_D - \mu_C < 0$, where H_a is called the alternate hypothesis. In this case, the entire α area is assigned to a single tail of the test distribution. See Figure 4.4. The α for a decrease in values is located in the lower tail since H_0 is contradicted for values of $(\bar{Y}_D - \bar{Y}_C) < 0$. The α for an increase would be located in the upper tail.

The difference between one-tailed and two-tailed hypothesis tests lies not in the calculational steps but rather in the determination of the rejection region and in the conclusions to be drawn. The decision whether to use a

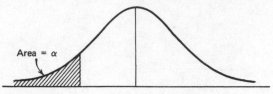

Figure 4.4. One-tailed rejection region.

one-tailed or two-tailed test depends on the alternate hypothesis of interest to the researcher. Such statements usually contain the phrases "less than" or "greater than" when comparing two populations.

One-sided tests can be used with both the z and t statistics.

METHODOLOGY

Inferences Concerning Single Samples

Example of a single sample z-test. An investigator was interested in determining the effect of a cardio-selective beta adrenergic blocking drug. Each of 25 healthy males, 24 to 38 years old, was given a single oral dose of 200 mg of the drug under study. Two hours later their pulse rates in beats/minute were taken, and the mean pulse rate for the group was found to be 68 beats/minute. The clinically accepted mean pulse rate for normal males is 72 beats/minute with a standard deviation of 10 beats/minute. Using the 5 percent level of significance, can it be concluded that pulse rate is the same for males on the medication and for normal males not taking the drug?

STEP 1. Statement of hypotheses. The null hypothesis is that the mean pulse rate for males taking the drug is the same as that for males not taking the drug. Symbolically,

$H_0: \mu = 72$

The alternate hypothesis is

$H_a: \mu \neq 72$

that is, the mean pulse rate for males on the drug is not the same as the mean pulse rate for males not taking the drug.

STEP 2. The following values are determined from the sample (or assumed to be known):

$$\bar{Y} = 68 \qquad \sigma/\sqrt{n} = \frac{10}{\sqrt{25}} = 2.0$$

$$n = 25$$

STEP 3.   The value of σ is assumed to be known, therefore the z test
is appropriate.

$$z = \frac{\bar{Y} - \mu}{\sigma/\sqrt{n}} = \frac{68 - 72}{2.0} = -2.0$$

STEP 4.   Determination of rejection region: for the 5 percent level of
significance, $\alpha = 0.05$, the appropriate critical z value from
Appendix Table B is $z_{.475} = \pm 1.96$.

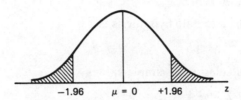

STEP 5.   Conclusion—Since the value calculated for z (step 3) falls in
the rejection region (shaded area, step 4), it is unlikely that the
discrepancy between the observed value ($\bar{Y} = 68$) and the
hypothesized value ($\mu = 72$) occurred by chance. Therefore,
the null hypothesis is rejected at the $\alpha = 0.05$ level of significance and it may be concluded that pulse rate for males on
the medication is different from that of normal males not on
the medication.

CI on μ.   A 95 percent confidence interval on the true mean pulse rate for
males on the experimental drug is given by

$$CI_\mu = \bar{Y} \pm 1.96\sigma/\sqrt{n}$$

$$= 68 \pm 1.96\,\frac{10}{\sqrt{25}}$$

$$= [64.1, 71.9]$$

Interpretation—We are 95 percent confident that the interval $[64.1, 71.9]$
covers the true mean pulse rate for all males on the drug.

Example of a single sample t-test.   A study was carried out to determine the
effect, if any, of a certain disease on plasma potassium levels. Determinations
were made on 10 individuals with the disease, and their mean plasma
potassium level was found to be 3.4 mEq/liter. The standard deviation for
this sample of measurements was 0.50 mEq/liter. The mean plasma potassium

level for normal individuals is assumed to be 4.5 mEq/liter. At the 0.01 level of significance, can it be concluded that plasma potassium level is different for patients with the disease?

STEP 1. Statement of hypotheses. The null hypothesis is that the mean plasma potassium level for the disease group is the same as the normal value, (i.e., plasma potassium levels are not affected by the disease). Symbolically,

$$H_0:\mu = 4.5$$

The alternate hypothesis is

$$H_a:\mu \neq 4.5$$

STEP 2. Calculation of sample values

$$\bar{Y} = 3.4 \qquad s = 0.50$$

$$n = 10$$

$$s/\sqrt{n} = \frac{0.50}{\sqrt{10}} = 0.16$$

STEP 3. The value of σ, the true population standard deviation, is *unknown*, therefore the *t*-test is appropriate.

$$t = \frac{\bar{Y} - \mu}{s/\sqrt{n}} = \frac{3.4 - 4.5}{0.16} = -6.9$$

STEP 4. Determination of rejection region. The critical value of t for $\alpha = 0.01$ and $df = n - 1 = 9$ is $t_{.005} = \pm 3.25$.

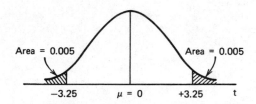

STEP 5. Conclusion—The calculated t value falls in the rejection region (shaded area); chance (sampling error) is therefore an unlikely explanation for the difference between the observed mean potassium level ($\bar{Y} = 3.4$) and the hypothesized value ($\mu = 4.5$). The null hypothesis is rejected at the 0.01 level of significance, and it may be concluded that plasma potassium level is different for patients with the disease.

CI on μ using t.   The 99 percent confidence interval on the true mean plasma potassium level for patients with the disease is given by

$$CI_\mu = \bar{Y} \pm t_{.005}s/\sqrt{n}$$

$$= 3.4 \pm 3.25 \frac{0.50}{\sqrt{10}}$$

$$= [2.9, 3.9]$$

Interpretation—We are therefore 99 percent confident that the interval [2.9, 3.9] covers the true mean plasma potassium level of patients with the disease.

Inferences Concerning Difference Between Two Means

Example using z.   In an experiment to determine the effect of exercise on the production of a certain substance P in the body, 200 normal males aged 20–35 were selected; 100 were randomly assigned to the excercise group and 100 were randomly assigned to the control group. The individuals in the treatment (exercise) group were required to spend 10 minutes on an exercise apparatus, rest 5 minutes, and continue on the apparatus. At the end of 60 minutes, the procedure was terminated and measurements of the substance P were made. The mean level of substance P for the exercise group was 0.58 ng/ml and the mean level of substance P for the control group was 0.53 ng/ml. Assume that $\sigma_1 = \sigma_2 = 0.28$. At the $\alpha = 0.05$ level of significance, can it be concluded that the production of substance P in the body is different for the exercise and control groups?

STEP 1.   Statement of hypotheses. The null hypothesis states that the mean level of substance P in the body is the same for the treatment and control groups (i.e., exercise does not affect the production of substance P). Symbolically,

$$H_0 : \mu_1 = \mu_2, \text{ or } \mu_1 - \mu_2 = 0$$

The alternate hypothesis is

$$H_a : \mu_1 \neq \mu_2, \text{ or } \mu_1 - \mu_2 \neq 0$$

STEP 2.   Calculation of sample values

Exercise Group	Control Group
$n_1 = 100$	$n_2 = 100$
$\bar{Y}_1 = 0.58$	$\bar{Y}_2 = 0.53$

STEP 3.   Calculation of test statistic. In this example, σ is known, therefore z is the appropriate test.

$$z = \frac{(\bar{Y}_1 - \bar{Y}_2)}{\sigma \sqrt{\dfrac{1}{n_1} + \dfrac{1}{n_2}}} = \frac{0.58 - 0.53}{0.28 \sqrt{\dfrac{1}{100} + \dfrac{1}{100}}} = 1.28$$

STEP 4.   Determination of rejection region. For $\alpha = 0.05$ level of significance, the critical value is $z_{.475} = \pm 1.96$.

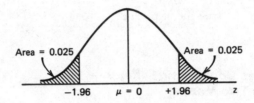

STEP 5.   Conclusion—The calculated z value (step 3) does not fall into the rejection region, therefore chance (sampling error) is a likely explanation for the observed difference between the treatment and control groups. It is concluded that there is insufficient evidence to reject the null hypothesis (at the 0.05 level of significance). The data do not show that the production of substance P in the body is different for the two groups.

CI on ($\mu_1 - \mu_2$).   A 95 percent confidence interval on the true mean difference in level of substance P for the exercise and control groups is given by

$$CI_{(\mu_1 - \mu_2)} = (\bar{Y}_1 - \bar{Y}_2) \pm z_{(.475)} s_p \sqrt{\frac{1}{n_1} + \frac{1}{n_2}}$$

$$= (0.58 - 0.53) \pm 1.96(0.039)$$

$$= [-.026, .126]$$

Interpretation—We are 95 percent confident that the interval $[-.026, .126]$ covers the true mean difference in level of P for the exercise and control groups.

Example using pooled *t*-test.   In order to evaluate the difference in serum Na levels between normotensive and newly diagnosed hypertensive patients not yet on a Na controlled diet, the following data were obtained:

	n	Mean (mEq/liter)	Standard Deviation (mEq/liter)
Normotensive	15	144	6.2
Hypertensive	12	160	3.9

Using an $\alpha = 0.10$ level of significance, can it be concluded that there is a difference in Na level for the normotensive and hypertensive groups?

STEP 1. Statement of hypotheses. The null hypothesis is that the mean Na level is the same for the normotensive and hypertensive groups. Symbolically,

$$H_0 : \mu_1 = \mu_2, \text{ or } \mu_1 - \mu_2 = 0$$

The alternate hypothesis is

$$H_a : \mu_1 \neq \mu_2, \text{ or } \mu_1 - \mu_2 \neq 0$$

STEP 2. Calculation of sample values

Hypertensive Group	Normotensive Group
$n_1 = 12$	$n_2 = 15$
$\bar{Y}_1 = 160$	$\bar{Y}_2 = 144$
$s_1 = 3.9$	$s_2 = 6.2$

$$s_p = \sqrt{\frac{(n_1 - 1)s_1^2 + (n_2 - 1)s_2^2}{n_1 + n_2 - 2}}$$

$$= \sqrt{\frac{11(3.9)^2 + 14(6.2)^2}{12 + 15 - 2}} = 5.3$$

$$s_p \sqrt{\frac{1}{n_1} + \frac{1}{n_2}} = 5.3 \sqrt{\frac{1}{12} + \frac{1}{15}} = 2.05$$

STEP 3. Calculation of test statistic. Since σ is unknown, the t test is appropriate. Further, we have two independent samples randomly selected from the two groups; therefore the pooled t rather than the paired t should be employed.

$$t = \frac{\bar{Y}_1 - \bar{Y}_2}{s_p \sqrt{\frac{1}{n_1} + \frac{1}{n_2}}} = \frac{160 - 144}{2.05} = 7.80$$

STEP 4. Determination of rejection region for $\alpha = 0.10$. The critical t value from Table C in the Appendix for $n_1 + n_2 - 2 = 25df$ is $t_{.05} = \pm 1.708$.

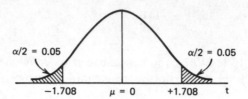

STEP 5. Conclusion—The calculated t value (step 3) falls in the rejection region and therefore chance (sampling error) is an unlikely explanation for the observed difference between Na levels for the normotensive and hypertensive groups. The null hypothesis is rejected at the $\alpha = 0.10$ level of significance, and it is concluded that serum Na level is different for the two groups.

CI on $\mu_1 - \mu_2$. The 90 percent confidence interval on the true mean difference in Na levels for the normotensive and hypertensive groups is

$$
\begin{aligned}
\mathrm{CI}_{(\mu_1 - \mu_2)} &= (\bar{Y}_1 - \bar{Y}_2) \pm t_{.05} s_p \sqrt{\frac{1}{n_1} + \frac{1}{n_2}} \\
&= (160 - 144) \pm 1.708(2.05) \\
&= [12.5, 19.5]
\end{aligned}
$$

Interpretation—We are 90 percent confident that the interval [12.5, 19.5] covers the true difference in the Na level for the two groups. Thus, the hypertensive group has a mean Na level at least 12.5 mEq/liter higher than the normotensive group at $\alpha = 0.10$.

The paired t-test. In a study on the efficacy of zinc sulfate for improving the physiological, psychological, and clinical status of geriatric patients with senile dementia, five geriatric patients were treated with 200 mg of the drug three times daily for 10 weeks. A control group of eight geriatric patients with diagnosed senile dementia received an identically appearing placebo administered under similar conditions. In conjunction with other measures, each member of both the treatment and the control groups was weighed prior to the start of the treatment period and again upon termination of the medication. The following tabulation gives before and after weight measurements for the treatment group. The investigator wished to determine if administration of zinc sulfate in geriatric patients affected the weight of the patients.

	Weight (lb)		
Patient	Before	After	Differences
1	140	143	+3
2	138	136	−2
3	142	138	−4
4	130	125	−5
5	152	150	−2

$$\sum d_i = -10$$
$$\sum d_i^2 = 58$$

Note: The order of subtraction is not important as long as all subtractions are carried out consistently and the formulation of the problem reflects the appropriate order. In this example, the order is "after" minus "before." Notice the corresponding order in the remainder of the problem.

STEP 1. Formulation of the hypotheses. Null hypothesis: The administration of oral zinc sulfate does not affect weight (lb) level in geriatric patients, that is, there is no difference in the before and after weight levels for the patients receiving the medication. Symbolically,

$$H_0 : D = 0$$

where D represents the true population mean of the distribution of differences. Alternate hypothesis: There is a statistically significant difference in the before and after weight levels for the patients receiving the medication. Symbolically,

$$H_a : D \neq 0.$$

STEP 2. Calculation of sample values

$$\bar{d} = \frac{\sum d_i}{5} = -2.0$$

$$s_d = \sqrt{\frac{\sum d_i^2 - \frac{(\sum d_i)^2}{5}}{4}} = 3.08$$

$$s_d / \sqrt{n} = \frac{3.08}{\sqrt{5}} = 1.38$$

STEP 3. Calculation of test statistic

$$t = \frac{\bar{d} - D}{s_d/\sqrt{n}} = \frac{-2.0}{1.38} = -1.45$$

STEP 4. Determination of rejection region. Using an $\alpha = 0.05$ level of significance, the critical t value for $n - 1 = 4df$ from Table C is $t_{.025} = \pm 2.776$.

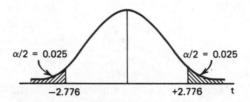

STEP 5. Conclusion—The calculated t value does not fall into the rejection region; chance is a likely explanation for the discrepancy between the after and before weight levels for the treatment group. The null hypothesis of no difference in weight levels cannot be rejected at the $\alpha = 0.05$ level of significance. Therefore, it cannot be concluded that there exists a statistically significant difference in the before and after weight levels for the patients receiving the medication.

CI on D. The 99 percent confidence interval on the true mean difference in the before and after weight levels is given by

$$\bar{d} \pm t_{.005} \frac{s_d}{\sqrt{n}}$$

$$-2.0 \pm 4.604(1.38) = [-8.4, 4.4]$$

Interpretation—We are 99 percent confident that the interval $[-8.4, 4.4]$ covers the true mean of the differences in weight levels before and after the administration of zinc sulfate.

Example of One-sided Tests of Hypothesis

Example using single sample t-test. Refer to the example of a single sample t-test given on page 93. Suppose it is of interest to determine whether the patients with the disease have lower plasma potassium than normal individuals. As stated in the section on one-sided test of hypothesis, the computational steps for the one-tailed and two-tailed tests are identical; the difference is in the statement of hypotheses to be tested and in the determination of the rejection region for the test.

The hypotheses to be tested in this example are:

$$H_0: \mu = 4.5$$

$$H_a: \mu < 4.5$$

where the alternate hypothesis is that the mean plasma potassium level for individuals with the disease is *lower than* 4.5.

For this example, α is 0.01 and the rejection region is the corresponding shaded area in the lower tail as shown below:

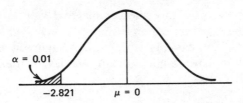

The critical t value from Table C corresponding to an area of 0.01 and $df = 9$ is

$$t_{.01} = -2.821$$

Note: For critical t values corresponding to areas in the lower tail, obtain the value corresponding to t_α and affix a minus sign.

Since the calculated t value ($t = -6.9$, step 3) falls in the rejection region, it may be concluded at the $\alpha = 0.01$ level of significance that patients with the disease have lower plasma potassium levels than normal individuals.

Example using pooled *t*-test. Refer to the example using pooled t-test given on page 96. Test the hypothesis that persons with essential hypertension have elevated serum Na levels. (Perform steps 2 and 3 as before.)

STEP 1. Statement of hypotheses.

$$H_0: \mu_1 - \mu_2 = 0$$

$$H_a: \mu_1 - \mu_2 > 0$$

STEP 4. Determination of rejection region. The rejection region for the example has the entire area $\alpha = 0.10$ in the upper tail as shown below:

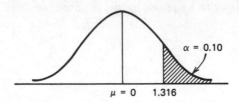

The critical t value $\alpha = 0.10$ and $25df$ is

$$t_{.10} = +1.316$$

STEP 5. Conclusion—Since the calculated t value ($t = +7.80$) falls in the rejection region it may be concluded (at the $\alpha = 0.10$ level) that patients with newly diagnosed essential hypertension have higher serum Na levels than normotensive individuals.

Statistical Decisions and Their Outcomes

Up to this point, the discussion of tests of hypotheses has supposed a preselected value of α, the level of significance of the test. In statistical decision making, the decision maker has the opportunity to select α. In doing so, he or she is determining how often an observation will be called unusual when in fact it is not unusual at all. Specifically, when a value for α, say 5 percent, is chosen, then in the long run, if the null hypothesis is true, it will be falsely rejected 5 percent of the time. For example, suppose we are testing the simple hypothesis

$$H_0 : \mu = \mu_0$$

versus

$$H_a : \mu > \mu_0$$

where μ is the mean of a normally distributed population and σ^2 is known. The decision rule for the test for $\alpha = .05$ calls for rejection of the null hypothesis when the test statistic $z = (\bar{Y} - \mu_0)/(\sigma/\sqrt{n})$ is greater than 1.65 (Fig. 4.5).

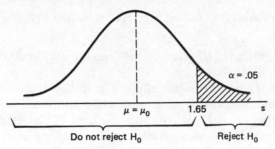

Figure 4.5.   Rejection region for $\alpha = .05$ (one-sided test)

When H_0 is true ($\mu = \mu_0$), we would expect 5 percent of the sample means ($\bar{Y}$), and, hence, 5 percent of the values of the test statistic, to fall in the rejection region (shaded area) for the test. Thus, we would expect to reject H_0 5 percent of the time when H_0 is really true. This type of incorrect decision, that is, rejecting a true H_0, is called a *Type I error*, and the probability of committing a Type I error is α. We have

$$\Pr(\text{reject } H_0 | H_0 \text{ true}) = \Pr(\text{Type I error}) = \alpha$$

The above is read "the probability that we reject H_0 given that H_0 is true" is α. An important concept that must be remembered is that we can never know the true "state of nature," that is, we cannot know whether H_0 is true or false. What we are saying is that "if H_0 is true," we risk falsely rejecting it (α) $\times$ 100 percent of the time.

If the null hypothesis is false and μ is not equal to μ_0 but rather is equal to some other value, say μ_1, then we make an incorrect decision, called a *Type II error*, if we fail to reject H_0. Consider Table 4.1. We see from this

Table 4.1.   Outcomes of statistical decisions

	True State of Nature	
	Data are from a population for which	
Statistical Decision	H_0 *True*	H_0 *False and* H_a *True*
Do not reject H_0	Correct decision	Incorrect decision: Type II error $\Pr(\text{Type II error}) = \beta$
Reject H_0	Incorrect decision: Type I error $\Pr(\text{Type I error}) = \alpha$	Correct decision

table that the probability of failing to reject a false null hypothesis is denoted by β.

$$\text{Pr(do not reject } H_0 | H_0 \text{ false)} = \text{Pr(Type II error)} = \beta$$

This probability is described pictorially in Figure 4.6b. If the null hypothesis is true, we are sampling from a population whose true mean is μ_0. The probability of rejecting H_0 if H_0 is true is α, the area of the shaded region in Figure 4.6a. If the null hypothesis is false, we are sampling from a population whose true mean is not μ_0 but some other value μ_1. The test of hypothesis is carried out, however, using μ_0 only. That is, the decision is made to reject H_0 if the value of the test statistic falls in the rejection region (shaded area) of Figure 4.6a. We do not reject H_0 if the test statistic falls in the nonrejection region (nonshaded area) of Figure 4.6a. Thus the probability of not rejecting H_0 when H_0 is true is the area of the nonshaded region in Figure 4.6a. However, if H_0 is *not true* and we are sampling from a population whose mean is μ_1, then the probability of not rejecting H_0 is given by the shaded area in Figure 4.6b.

If we are able to specify an exact value for μ_1 in the alternate hypothesis, then, for a specified α level, we can calculate β, the probability of a Type II error. However, in practice, we usually test hypotheses of the form

$$H_0 : \mu = \mu_0 .$$

$$H_a : \mu > \mu_0 \quad \text{or} \quad H_a : \mu < \mu_0 \quad \text{or} \quad H_a : \mu \neq \mu_0$$

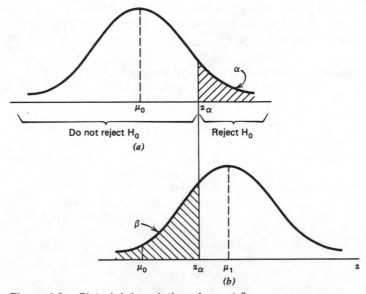

Figure 4.6. Pictorial description of α and β.

where the exact value of μ under H_a is not specified. In this case we cannot determine an exact value for β. We do know the following:

1. For a fixed sample size n, as α becomes smaller, β becomes larger. In Figure 4.7 we have decreased α from .05 to .01. As can be seen, as α decreases, the cutoff point for the rejection region in Figure 4.7a moves farther to the right. Correspondingly, the area of the shaded region in Figure 4.7b (β) becomes larger.

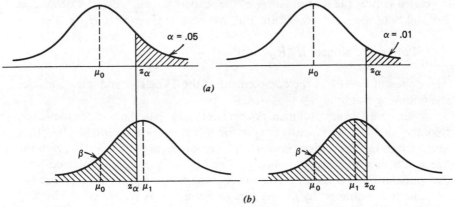

Figure 4.7. The size of α and β for α decreased to .01 from .05 (one-sided test).

2. As the difference between μ_0 and μ_1 increases, β decreases (Fig. 4.8).

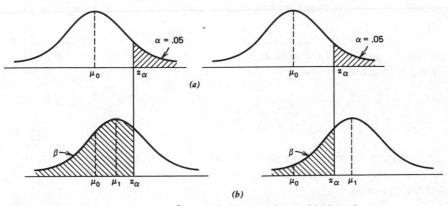

Figure 4.8. The size of α and β for μ_1 increased (one-sided test).

3. For a fixed α, as the sample size is increased, β is decreased, since the standard error $\sigma/\sqrt{n}$ decreases as n increases. The following discussion of the dependence of power on sample size will illustrate this point.

Power of the Test and Determination of Sample Size

In the preceding section we considered the probabilities of the two types of errors that can be made when making a statistical decision regarding the credibility of the null hypothesis. We stated

$$\Pr(\text{reject } H_0 | H_0 \text{ true}) = \alpha$$

$$\Pr(\text{do not reject } H_0 | H_0 \text{ false}) = \beta$$

A related concept is that of *power* of the statistical test. Power is defined as the ability of the test to reject the null hypothesis given that H_0 is false.

$$\text{Power} = \Pr(\text{reject } H_0 | H_0 \text{ false}) = 1 - \beta$$

The power of the test is the complement of the Type II error rate. Consider the following example.

Suppose a group of clinical researchers know, based on prior experience, that the mean of a clinical variable for a certain population is 50 with a known standard deviation, σ, of 15. The researchers plan to investigate the effect of a new treatment applied to a sample of individuals selected from this population. Further, they feel that if the treatment would increase the variable being studied by an average of 5 units, this would be a clinically meaningful result. Thus, they wish to test the hypothesis $H_0 : \mu = 50$ versus $H_a : \mu = 55$. How does the choice of sample size affect the probability of finding such an increase if the treatment is truly effective? In statistical terminology, what is the probability of rejecting the null hypothesis when the alternative hypothesis is true; that is, what is the power of the statistical test?

Suppose the researchers choose a sample of size $n = 9$ and for these nine individuals find the sample mean $\bar{Y}$ to be equal to 56. Thus, the value of the test statistic is

$$z = \frac{\bar{Y} - \mu_0}{\sigma/\sqrt{n}} = \frac{56 - 50}{15/\sqrt{9}} = 1.2$$

The rejection region for this test is given by the shaded area shown in Figure 4.9. Note that a value of 2.33 on the z scale corresponds to a value on the $\bar{Y}$ scale given by

$$\bar{Y} = z(\sigma/\sqrt{n}) + \mu_0 = 2.33(15/\sqrt{9}) + 50 = 61.65$$

Therefore, the decision rule for this test is to reject H_0 when the test statistic z is greater than 2.33. Correspondingly, in terms of $\bar{Y}$, we reject H_0 when the value of the sample mean $\bar{Y}$ is greater than 61.65. In this example,

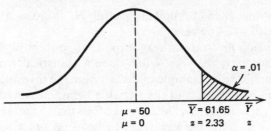

Figure 4.9. Rejection region for $\alpha = .01$.

the test statistic (or $\bar{Y}$) falls in the nonrejection region. We cannot reject H_0 and therefore cannot conclude that $\mu = 55$.

The power of the test for this example is computed as follows:

$$\beta = \text{Pr(do not reject } H_0 | H_0 \text{ false and } H_a \text{ true)}$$

$$= \text{Pr}(\bar{Y} \leq 61.65 | \mu = 55)$$

$$= \text{Pr}\left(\frac{\bar{Y} - \mu}{\sigma/\sqrt{n}} \leq \frac{61.65 - 55}{15/\sqrt{9}}\right)$$

$$= \text{Pr}(z \leq 1.33) = .9082$$

Power $= 1 - \beta = .0918$

Thus, with a sample of size 9, we have only a 9.18 percent chance of being able to reject a false null hypothesis. That is, if the true difference in means is as large as 5 units, with this particular test we have only a 9.18 percent chance of detecting such a difference.

If the study were done using $n = 100$ rather than $n = 9$, the reader may verify that the decision rule for $\alpha = .01$ is to reject H_0 when $\bar{Y}$ is greater than 53.495. Then

$$\beta = \text{Pr(do not reject } H_0 | H_0 \text{ false and } H_a \text{ true)}$$

$$= \text{Pr}(\bar{Y} \leq 53.495 | \mu = 55)$$

$$= \text{Pr}\left(z \leq \frac{53.495 - 55}{15/\sqrt{100}}\right) = \text{Pr}(z < -1.00)$$

$$= .1587$$

and Power $= 1 - \beta = .8413$.

By increasing the sample size from $n = 9$ to $n = 100$, we have increased the power of the test from 9.18 percent to 84.13 percent. The probability of detecting a true difference of 5 units (rejecting H_0 in favor of H_a when H_a is true) is increased to 84 percent.

This discussion illustrates a very important concept in the interpretation of statistical tests; that is, "What does a statistically nonsignificant result really mean?" In the example of this section, the investigator feels that a difference of at least 5 units in the variable being measured would be of clinical or practical importance. Differences of less than 5 units are considered by those familiar with the clinical issues to be of no practical importance. Therefore, it is desirable to construct a test which we feel, with reasonable probability, can detect a difference of at least 5 units, if a difference of this magnitude really exists. Assume, in the first example, that $H_a : \mu = 55$ is true; that is, a true difference of 5 units does exist. Based on the data, however, we are unable to reject the null hypothesis $H_0 : \mu = 50$ since $z_{cal} = 1.2$ is less than the critical value 2.33. How do we interpret this negative result? Many investigators are tempted to interpret a statistically nonsignificant result as demonstrating that the null hypothesis is "true." They proceed to declare that a difference of 5 units *does not exist* in the population. Consider, however, that the power of this test, when the sample consists of 9 observations, is only .0918. Therefore, even if a difference as large as 5 units does exist in the population, we have only a 9.18 percent chance of detecting it. To declare, then, that a difference of 5 units does not exist, based on the outcome of the statistical test, would indeed be misleading.

In the example with the sample size increased to $n = 100$, the power of the test has been increased to 84 percent. With this power, we know that if a difference of 5 units exists, then there is an 84 percent chance that the test will detect it. An investigator in this situation can have more faith that a statistically nonsignificant result does in fact suggest that a difference as large as 5 units does not exist in the population.

One important use of the idea of power of a statistical test is that of determination of a sample size, in advance, that will ensure a specified power for the test. The calculation of sample size, based on power considerations, requires that the investigator specify the following:

1. The size of the effect that is clinically worthwhile to detect
2. The probability of falsely rejecting a true null hypothesis (α)
3. The probability of failing to reject a false null hypothesis (β)
4. The standard deviation of the population being studied

The first three items above are under the control of the investigator. A value for the population standard deviation may be determined from previous work, work reported in the literature, a pilot study, or simply an educated guess.

For the one and two sample t-tests described in this chapter, Table F in the Appendix may be used to determine the sample size n for a given α, β, and Δ, where

$$\Delta = \frac{\mu - \mu_0}{\sigma}$$

Δ for the one sample case is the difference in units of standard deviation between the hypothesized and the true population means.

Example. Consider the example of the single sample t-test given on page 93. Suppose the investigator wishes to detect, with 80 percent certainty, a difference of one standard deviation between the hypothesized and the true mean plasma potassium level of the diseased population. We have

$$\alpha = .01$$
$$\text{Power} = .80$$
$$\beta = 1 - .80 = .2$$
$$\Delta = 1$$

From Table F, we find that for the one-sided test $(H_a : \mu < 4.5)$ for $\alpha = .01$, the sample size must be equal to 13. If the investigator wishes to detect a difference as small as 0.30 standard deviations with $\alpha = .01$ and Power $= .80$, then the sample size must be increased to at least 115 observations. By increasing the sample size from 13 to 25 for $\alpha = .01$, the power of the test to detect a difference of one standard deviation is increased from .80 $(\beta = .2)$ to .99 $(\beta = .01)$.

Example. Refer to the example of the pooled t-test given on page 96. Suppose the investigator wishes to detect, with 90 percent certainty, a difference of one standard deviation, if it exists, in the mean serum Na levels of the normotensive and the newly diagnosed hypertensive populations. From Table F, for $\alpha = .10$ (two-sided test), Power $= .90$, $\beta = 1 - .90 = .10$, and $\Delta = 1$, we find that $n_1 = n_2 = 18$. The sample size must be increased to $n_1 = n_2 = 70$ if we wish to detect a difference in population means as small as 0.5 standard deviations. Thus, as the difference in population means decreases, the required sample sizes increase sharply.

p Values

Often, researchers report the exact probability p of obtaining a sample value equal to or more extreme than that observed, assuming the null hypothesis is true. For example, in the problem for the single sample z-test given on page 92, the probability of obtaining a sample mean equal to or smaller than 68 from a population whose true mean is 72 is the shaded area depicted below:

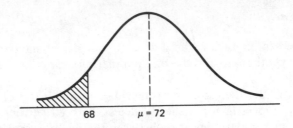

$$68 \qquad \mu = 72$$

We wish to find

$$\Pr(\bar{Y} \leq 68 | \mu = 72)$$

where the above statement is read "the probability that the sample mean $\bar{Y}$ is less than or equal to 68 *given that* the true population mean μ is 72."

$$\Pr(\bar{Y} \leq 68 | \mu = 72) = \Pr\left(\frac{\bar{Y} - \mu}{\sigma/\sqrt{n}} \leq \frac{68 - 72}{2}\right) = \Pr(z \leq -2)$$

$$= .5 - .4772$$

$$= .0228$$

$$= p$$

That is, if the null hypothesis, $\mu = 72$, is true, the probability of obtaining a test statistic z as small or smaller than -2 is .0228. Note that this p value is appropriate when we are interested in detecting the one-tailed alternative $\mu < 72$. To obtain the p value for the two-tailed alternative, $\mu \neq 72$, we double the p value to obtain p(two-tailed) $= 2(.0228) = .0456$. If the exact p value is less than the predetermined α, then the null hypothesis is rejected.

Owing to the lack of necessary tables, it is sometimes difficult to compute the exact value of p. For example, when we are using the t distribution rather than the z distribution, as is done in the example on page 97, we cannot determine from Table C in the Appendix the exact probability of obtaining a test statistic as large or larger than $t = 7.8$ given that the null hypothesis $H_0: \mu_1 = \mu_2$ is true. It can be determined from Table C, however, that for $df = 25$, the probability of observing a t value as large or larger than 2.7874 is .005. Since the actual test statistic ($t = 7.8$) is larger than the tabulated value for $\alpha = .005$, it is obvious that the probability of obtaining a test statistic of 7.8 is smaller than .005. Thus, we report for this example that $p < .005$ for the one-tailed case or $p < .01$ for the two-tailed case. If, in a similar example, the value of the test statistic had been $t = 2.3$, we see from Table C for $df = 25$ that $p < .025$ for the one-sided test or $p < .05$ for the two-sided test. The p value for a test is always the smallest value of α for which the null hypothesis can be rejected.

Statistical Significance Versus Practical or Clinical Significance

An important concept in the interpretation of a statistical test is that of statistical significance versus practical significance. Consider an example in which an investigator is comparing the mean systolic blood pressure level of a group of patients being given an experimental antihypertensive drug with the mean systolic blood pressure of a control group. For the purpose of illustration, suppose that the true difference in the mean systolic blood pressure of the populations from which the two groups were selected is 1 mmHg $(\mu_T - \mu_C = 1)$. As described earlier, we may choose the sample size n so that the statistical test has sufficient power to detect even this small a difference in population means. Thus, while we may declare a "statistically significant" difference in the mean blood pressure of the control and treatment groups, we may be unwilling to promote the use of a drug which, on the average, may reduce blood pressure by only 1 mmHg. The point to remember is that statistical significance does not imply necessarily that the true difference in population means is of sufficient magnitude to be of clinical or practical importance. For large sample sizes, the actual difference in the observed sample values $(\bar{Y}_1 - \bar{Y}_2)$ is usually a good estimate of the actual difference in the means $(\mu_1 - \mu_2)$ of the populations being studied. However, the decision as to whether a difference is of practical importance is not a statistical one, but rather must be made by those with expertise in the subject area being investigated.

PROBLEMS

1. Suppose the mean urine chloride level in normal full-term infants is assumed to be 210 mEq/24 hours and the standard deviation is 20 mEq/24 hours. The mean urine chloride for 25 premature infants was found to be 170 mEq/24 hours. Based on this data, can it be concluded that premature infants have a lower urine chloride level than the clinical norm? (Use $\alpha = 0.01$.)

2. Place a 99 percent confidence interval on the true mean urine chloride level for premature infants.

3. For nine adult male federal prisoners addicted to heroin the mean glucose-6-dehydrogenase level in the blood was found to be 450 units/10^9 cells and the standard deviation (s) was 62.5 units/10^9 cells. The adult normal value for glucose-6-dehydrogenase is 375 units/10^9 cells. Do the data present sufficient evidence to suggest that production of glucose-6-dehydrogenase in the blood is different for heroin addicts? (Use $\alpha = 0.05$.)

4. Place a 95 percent confidence interval on the true mean level of glucose-6-dehydrogenase for heroin addicts. How would you interpret this interval?

5. The following tabulation presents the mean cumulative weight loss in grams for 12 patients receiving propranolol and for 11 control patients following sweating during insulin-induced hypoglycemia.

	n	Mean Weight Loss (in grams)	Standard Deviation
Propranolol	12	120	10.0
Control	11	70	8.0

Do the data present sufficient evidence to conclude that the mean cumulative weight loss is different for the two groups? (Use $\alpha = 0.05$.)

6. For problem 5, place a 95 percent confidence interval on the true difference in mean cumulative weight loss following insulin-induced hypoglycemia for the propranolol and control groups.

7. Serum digoxin levels were determined for nine healthy males aged 20–45 following rapid intravenous injection of the drug. The measurements were made 4 hours after the injection and again at the end of an 8-hour period.

Serum Digoxin Concentration (ng/ml)

Subject	4 hours	8 hours
1	1.0	1.0
2	1.3	1.3
3	0.9	0.7
4	1.0	1.0
5	1.0	0.9
6	0.9	0.8
7	1.3	1.2
8	1.1	1.0
9	1.0	1.0

Is there a statistically significant difference (at the $\alpha = 0.05$ level) in the serum digoxin concentration at the end of 4 hours and the concentration at the end of 8 hours?

8. For the above problem, place a 95 percent confidence interval on the true mean difference in serum digoxin concentration for the 4-hour and 8-hour periods.

9. Refer to problem 7. Another group of 11 healthy males from the same age group were given digoxin by intravenous infusion. The mean serum digoxin concentration at the end of 8 hours was found to be 0.95 ng/ml and the standard deviation was 0.2 ng/ml. Can it be concluded that the serum digoxin concentration is higher 8 hours following rapid injection than at the end of 8 hours of intravenous infusion? (Use $\alpha = 0.01$.)

10. For problem 9, place a 99 percent confidence interval on the true mean difference in serum digoxin concentrations 8 hours following rapid injection and 8 hours following intravenous infusion.

11. The mean plasma potassium level for 50 adult males with a certain disease was found to be 3.35 mEq/liter and the standard deviation was 0.50 mEq/liter. The adult normal value for plasma potassium is 4.6 mEq/liter. Based on the above data, can it be concluded that males with the disease have lower plasma potassium levels than normal males? (Use $\alpha = 0.05$.)

12. Place a 95 percent confidence interval on the true mean plasma potassium level for males with the disease.

13. Refer to problem 6, Chapter 3. Based on the data, can it be concluded that mean urinary calcium excretion is different for the bumetanide and the control groups (Use $\alpha = 0.05$.)

14. Place a 95 percent confidence interval on the true difference in mean calcium excretion rate for the bumetanide and control groups.

15. Refer to problem 7, Chapter 3. Test the hypothesis that cholesterol measurements are no higher for patients who have suffered a coronary event than for normal healthy males against the alternative that patients with a coronary event have elevated serum cholesterol levels. (Use $\alpha = 0.05$.)

16. If a new analyzer for processing blood samples is to be put into full-time operation, it must be capable of analyzing on the average more than 100 samples per day. A 5-day trial period produces the following results:

Day	Number of Blood Samples Processed
1	101
2	90
3	94
4	98
5	102

Based on this data, would you purchase the new analyzer? (Use $\alpha = 0.05$.)

17. The serum indirect bilirubin levels were determined for five premature and for five full-term infants. If it is reasonable to assume that the two populations have equal standard deviations (i.e., $\sigma_1 = \sigma_2 = 1.0$ mg/100 cc), can it be concluded that the mean serum indirect bilirubin level is different for the two groups? (Use $\alpha = 0.01$.)

   Serum Indirect Bilirubin
   (mg/100 cc)

Premature	Full-term
1.0	2.0
2.0	4.0
3.0	6.0
2.0	4.0
2.0	4.0

18. Place a 99 percent confidence interval on the true difference in the mean serum indirect bilirubin levels for the premature and full-terms babies.

19. In order to determine the effect of a certain oral contraceptive on weight gain, nine healthy females were weighed prior to the start of the medication and again at the end of a 3-month period.

Subject	Initial	3 Months
1	120	123
2	141	143
3	130	140
4	150	145
5	135	140
6	140	143
7	120	118
8	140	141
9	130	132

   Is there sufficient evidence to conclude that females experience a weight gain following 3 months of the oral contraceptive use? (Use $\alpha = 0.05$.)

20. Place a 95 percent confidence interval on the true mean difference in weight prior to and following 3 months of the oral contraceptive use.

CHAPTER 5

REGRESSION
AND CORRELATION

OVERVIEW

In medical research it is often desirable to obtain a mathematical expression by which the value of one variable might be used to predict the value of another. For example, it may be of interest to the physician to be able to predict a patient's expected survival time following a heart attack on the basis of the patient's age, or he may wish to predict a patient's blood pressure for a given dose of an antihypertensive drug.

Also, a frequent aim of medical investigation is to determine whether a relationship exists between certain variables, and if so, to obtain a numerical measure of the "closeness" of the association between the variables. The current research effort directed toward defining the association between cigarette smoking and lung cancer is an illustration of this area of statistical analysis.

The above cases are examples of two widely used and closely associated techniques of statistical analysis: regression and correlation. This chapter presents methods for obtaining a prediction equation from observed data (regression analysis) and describes the procedure for measuring the degree of association between two observed variables (correlation analysis).

OBJECTIVES

- To be able to plot scatter diagrams
- To describe the relationship between two variables by finding a regression equation for the two variables
- To test hypotheses concerning the slope of the regression equation
- To place confidence intervals on the slope of the regression equation

- To measure the degree of a relationship between two variables by finding the correlation coefficient
- To list the precautions associated with the use of regression and correlation analyses

FOUNDATIONS

Previously, the samples studied have consisted of measurements on a single random variable, Y. When two random variables, X and Y, are measured, the techniques of *regression* and *correlation* are applicable.

Regression and correlation, while closely related, are used for different purposes. The most common objective of regression analysis is to obtain an equation that may be used to predict or estimate the value of one variable corresponding to a given value of the other variable. Correlation analysis, on the other hand, is used to obtain a measure of the degree, or strength of the association between two variables.

For both regression and correlation situations, data consist of pairs of measurements selected from the population of interest. For example, a medical school admissions committee may wish to investigate whether undergraduate college grade point average has any predictive association with medical school grades. The data could be arranged as shown below, where

Student	College GPA	Medical School GPA
1	X_1	Y_1
2	X_2	Y_2
⋮	⋮	⋮
n	X_n	Y_n

the pair of numbers (X_i, Y_i) give the college GPA (X_i) and the medical school GPA (Y_i) of the ith student in the sample.

As is the case in all statistical inference, a decision as to whether it is reasonable to assume that a relationship actually exists between Y and X has its basis in probability. A mathematical "equation" which defines the relationship between the dependent variable Y and the independent variable X is obtained from a sample and then a statistical test of hypothesis enables us to reach an objective decision. The equation can then be used to predict values of the dependent variable (Y) from values of the independent variable (X).

Before an equation relating X and Y can be estimated, we must assume the *functional form* of the relationship (e.g., the relationship may be linear, quadratic, or exponential). In any given situation, the choice of functional form may be determined in three ways: (1) from analytical or theoretical considerations, (2) from experience, or (3) by studying a scatter diagram.

The Scatter Diagram

The scatter diagram provides probably the simplest and most useful means for studying the relationship between two variables. In a scatter diagram each of the n pairs of observations (Y_i, X_i) is plotted as a single point on a graph. The Xs are plotted on the horizontal axis and the Ys are plotted on the vertical axis. By looking at the arrangement of points on the graph, one may be able to discern a pattern indicative of the nature of the underlying functional form of the data.

Some scatter plots are shown in Figure 5.1. Figure 5.1a suggests the existence of a linear relationship between the variables Y and X. A *linear* relationship is one that may be described by a straight line. Further, a positive linear relationship is suggested. For each increase in X there is a corresponding increase in Y. Conversely, for a negative linear relationship, the pattern of points would be downward to the right. In this case, as the value of X increased, the value of Y would decrease. Figure 5.1b represents a *curvilinear*, or nonlinear relationship. A straight line clearly does not represent the relationship. Figure 5.1c depicts a case of no relationship between variables. The value for Y, in this case, does not depend on the value of X.

In this chapter the discussion will be restricted to data that show a linear relationship.

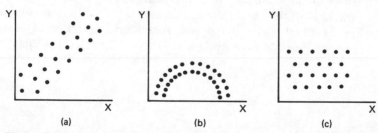

(a)   (b)   (c)

Figure 5.1.

Simple Linear Regression

As an introduction to the problem of regression analysis, data relating systolic blood pressure to dose level of an antihypertensive drug will be used.

Table 5.1 gives the systolic blood pressures for a group of hypertensive patients along with the dose level of a certain antihypertensive drug. Looking at the table, we can see that some relationship does exist. As the dose level increases, the mean systolic blood pressure decreases. These data are plotted in a scatter diagram in Figure 5.2. Observe that all the points do not fall exactly on a straight line but do seem to follow a uniform downward trend.

Table 5.1.

Dose Level of Drug (in mg)	Mean Systolic Blood Pressure (in mmHg)
2	278
3	240
4	198
5	132
6	111

This is an indication that the relationship could be linear and can be described by a straight line.

We now wish to determine the "equation" for the line shown in Figure 5.2. Recall that any straight line has the general form

$$Y = a + bX$$

where b is the slope of the line and a is the point where the line intercepts the Y axis. For any two points, it is a simple matter to determine the equation for the line connecting the two points. However, in statistical problems involving three or more sample points, it is generally impossible to find a line that goes through all the points simultaneously. In this case, we attempt to find the line that "best" fits all the points. A stylized illustration of data and such a prediction line drawn by eye are shown in Figure 5.3. The "hat" over

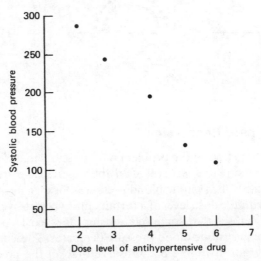

Figure 5.2.

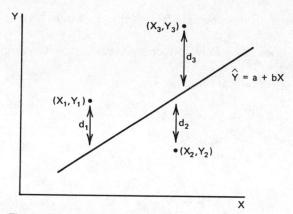

Figure 5.3.

Y, read "Y hat," in this figure is used to denote that this line is an estimate of some theoretically true line.

The distances of the observations from the line are given by

$$d_i = Y_i - \hat{Y}_i$$

where

$$\hat{Y}_i = a + bX_i$$

It seems to be a function of the human eye that such lines seem naturally to try to minimize the distances (or deviations) d_i of the observed data points Y_i from the line.

In Figure 5.3, d_1 is positive, d_2 is negative, and d_3 is positive. As before, deviations add to zero so it is more convenient to compute the sum of squared deviations and use the square root to measure the overall "fitness" of the line. That is, we will deal with

$$\sum d_i^2 = \sum (Y - \hat{Y})^2$$

Further, since different people would probably draw different lines, a standard statistical procedure is needed to find the line such that the sum of squared deviations is a minimum.

The statistical procedure for finding this best fitting line is called the *method of least squares* and the line is called the *regression line*. The formal derivation of this procedure, which requires differential calculus, is presented in advanced statistical texts. The results from the method of least squares will be applied to the data from Table 5.1 to illustrate the principles involved.

First, it is necessary to introduce some useful new notation:

$$(X_i, Y_i) = i\text{th pair of observations}$$

$$\sum_{i=1}^{n} (X_i - \bar{X})(Y_i - \bar{Y}) = \sum XY - \frac{(\sum X)(\sum Y)}{n} = \sum xy$$

$$\sum_{i=1}^{n} (Y_i - \bar{Y})^2 = \sum Y^2 - \frac{(\sum Y)^2}{n} = \sum y^2$$

$$\sum_{i=1}^{n} (X_i - \bar{X})^2 = \sum X^2 - \frac{(\sum X)^2}{n} = \sum x^2$$

The sample regression line is written

$$\hat{Y} = \hat{\beta}_0 + \hat{\beta}_1 X$$

where the least squares estimates $\hat{\beta}_0$ and $\hat{\beta}_1$ are (without proof)

$$\hat{\beta}_1 = \frac{\sum xy}{\sum x^2} \quad \text{and} \quad \hat{\beta}_0 = \bar{Y} - \hat{\beta}_1 \bar{X}$$

The values $\hat{\beta}_0$ and $\hat{\beta}_1$ are calculated from a sample of observations from the entire population of interest and are estimates of the "true" population values β_0 and β_1. As was the case with $\bar{Y}$ and s, the values $\hat{\beta}_0$ and $\hat{\beta}_1$ are subject to sampling variation and therefore may vary from sample to sample.

Table 5.2. Relationship between dose level of antihypertensive drug and systolic blood pressure

Dose Level (X)	Systolic Blood Pressure (Y)	X^2	Y^2	XY
2	278	4	77284	556
3	240	9	57600	720
4	198	16	39204	792
5	132	25	17424	660
6	111	36	12321	666
Total 20	959	90	203,833	3394

The data in Table 5.1 are rewritten in Table 5.2 together with the required computations and regression data.

From the totals in Table 5.2, we can now compute

$$\bar{Y} = \frac{\sum Y}{n} = \frac{959}{5} = 191.8 \qquad\qquad \bar{X} = \frac{\sum X}{n} = \frac{20}{5} = 4.0$$

$$\sum y^2 = \sum Y^2 - \frac{(\sum Y)^2}{n} \qquad\qquad \sum x^2 = \sum X^2 - \frac{(\sum X)^2}{n}$$

$$= 203{,}833 - \frac{(959)^2}{5} \qquad\qquad = 90 - \frac{(20)^2}{5}$$

$$= 19{,}896.8 \qquad\qquad\qquad\qquad = 10$$

$$\sum xy = \sum XY - \frac{(\sum X)(\sum Y)}{n}$$

$$= 3394 - \frac{(20)(959)}{5}$$

$$= -442$$

Thus,

$$\hat{\beta}_1 = \frac{-442}{10} = -44.2$$

$$\hat{\beta}_0 = 191.8 - (-44.2)4$$

$$= 368.6$$

and

$$\hat{Y} = 368.6 - 44.2X$$

Using the above regression equation, one may now predict the systolic blood pressure for a given dose of the drug for any patient on the medication. The values of $\hat{Y}$ for each X and the differences between the predicted values ($\hat{Y}$) and the observed values (Y) are shown in Table 5.3. The value $\hat{Y}$ obtained for a given X is the predicted mean of the population of all possible Y values that could occur at the given value X.

The estimated regression line is plotted in Figure 5.4. As can be noted from Table 5.3, the sum of the deviations of the observed points from the points predicted by the line is zero.

Table 5.3. Relationship between dose level of antihypertensive drug and systolic blood pressure

Values of $\hat{Y}$ for Each X	Difference Between Predicted $(\hat{Y})$ and Observed (Y) $(Y - \hat{Y})$
280.2	−2.2
236.0	4.0
191.8	6.2
147.6	−15.6
103.4	7.6

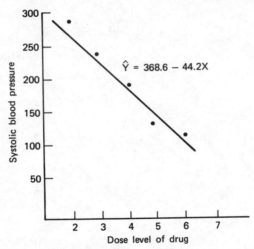

Figure 5.4. Regression of systolic blood pressure on drug dose level.

Variation About the Regression Line

Just as there is a sample standard deviation associated with each $\bar{Y}$, there is a standard deviation associated with the regression line and $\hat{Y}$. This quantity, denoted by $s_{y.x}$ to signify regression, is called the *standard error of the estimate*. It is given by

$$s_{y.x} = \sqrt{\text{SSE}/(n - 2)}$$

where n is the number of pairs of observations and SSE (sum of squares for error) is defined as

$$\text{SSE} = \sum (Y - \hat{Y})^2$$

The quantity $s_{y.x}$ is seen to be analogous to the standard deviation computed in previous chapters. It measures the "average" deviation of the observed values (Y) from the values $(\hat{Y})$ predicted by the regression line.

Although we will not test hypotheses or compute confidence intervals for an estimate $\hat{Y}$, the standard error (S.E.) for $\hat{Y}$ at a given X value would be

$$\text{S.E.} \ (\hat{Y}) = s_{y.x} \sqrt{\frac{1}{n} + (X - \bar{X})^2 / \textstyle\sum x^2}$$

From Table 5.3 we find

$$\text{SSE} = \sum (Y - \hat{Y})^2$$
$$= 360.4$$

and

$$s_{y.x} = \sqrt{360.4/3}$$
$$= 11.0$$

Fortunately, there is a computationally equivalent formula for SSE which is both more convenient to use and gives an insight into the geometric meaning of the regression line. This form is

$$\text{SSE} = \sum y^2 - \hat{\beta}_1 \sum xy$$
$$= 19,896.8 - (-44.2)(-442)$$
$$= 360.4$$

as before. From this computational form it is easily seen that the sum of squares of error is equal to the sum of squares of Y only if $\hat{\beta}_1 = 0$.

Also, the variation about the regression line, as measured by SSE, is strictly *less* than the Y variation, as measured by $\sum y^2$, whenever $\hat{\beta}_1 \neq 0$ (Note that $\hat{\beta}_1$ and $\sum xy$ always have the same algebraic sign so that the product $\hat{\beta}_1 \sum xy$ is always positive.) Consequently, whenever there is a linear relationship between X and Y we can compute a standard error based on this relationship which is smaller than the simple standard error based on Y values alone. It may be verified that the standard error of the Y values alone is $s_y/\sqrt{n} = 31.5$, which is considerably larger than $s_{y.x} = 11.0$.

Tests of Hypotheses Concerning β_1

The procedure for obtaining the least squares regression line is a mathematical one and can be applied to any set of data regardless of the underlying functional form. That is, the statistical procedure can be applied to data that resembles those shown in Figure 5.1c, although there is clearly no relationship between X and Y. It is for this reason that the sample regression line must be evaluated to determine if it adequately describes the relationship between the variables X and Y. This may be accomplished by testing the null hypothesis that the true slope β_1 of the population regression line is equal to zero (i.e., $H_0 : \beta_1 = 0$).

As stated previously, $\hat{\beta}_0$ and $\hat{\beta}_1$ are estimates of the true population parameters β_0 and β_1 and are subject to variation from sample to sample. It is important, therefore, to be able to make inferences concerning the true population regression line based on the information obtained from the sample regression line.

The most important inference to be made concerns the "true" value of the slope, β_1, of the population line. If the true population β_1 is zero, then the value of Y in no way depends on the value of X. In other words, a value of $\beta_1 = 0$ indicates that no linear relationship exists between X and Y.

As was the case with testing hypotheses about μ discussed in Chapter 4, it is first necessary to determine the standard error of $\hat{\beta}_1$ before any tests of significance on $\hat{\beta}_1$ can be performed. Without proof, the standard error of $\hat{\beta}_1$ is

$$\frac{s_{y.x}}{\sqrt{\sum x^2}}$$

The principles involved in tests of hypotheses and in constructing confidence intervals are exactly the same as those described in Chapter 4.

STEP 1.   Formulation of the null hypothesis.

$H_0 : \beta_1 =$ Some specified value (usually 0)

STEP 2.   Calculation of sample values

$$\hat{\beta}_1 = \frac{\sum xy}{\sum x^2} \quad \text{and} \quad \frac{s_{y.x}}{\sqrt{\sum x^2}}$$

STEP 3.   Calculation of test statistic

1. For $n > 30$, use

$$z = \frac{\hat{\beta}_1 - \beta_1}{s_{y.x}/\sqrt{\sum x^2}}$$

2. For $n \leq 30$, use

$$t = \frac{\hat{\beta}_1 - \beta_1}{s_{y.x}/\sqrt{\sum x^2}}, \quad n - 2 \text{ degrees of freedom}$$

STEP 4. Determination of rejection region as described in chapters 3 and 4.

STEP 5. Comparison of the calculated test statistic to the tabulated value to decide whether to reject or not reject the null hypothesis.

The most common form of the null hypothesis is $\beta_1 = 0$. If this null hypothesis is rejected, we conclude that there is a significant linear relationship between Y and X. If it is not rejected we must say that there is not enough evidence to show that β_1 differs significantly from zero and therefore must conclude that X does not contribute information for the prediction of Y.

The confidence interval on β_1 is given by

$$CI(\beta_1) = \hat{\beta}_1 \pm \theta_{\alpha/2} s_{y.x}/\sqrt{\sum x^2}$$

where θ is either z or t depending on the size of the sample.

For the data from Table 5.2 we have

$$\hat{\beta}_1 = -44.2, \qquad s_{y.x}/\sqrt{\sum x^2} = 3.5, \qquad n = 5$$

and for $H_0 : \beta_1 = 0$

$$t = \frac{-44.2}{3.5}$$

$$= -12.6, \text{ 3 degrees of freedom}$$

The tabulated t value for $\alpha = 0.05$ and 3 degrees of freedom is ± 3.182. Thus, we conclude that there is a significant negative linear relationship between systolic blood pressure and dose level of the antihypertensive drug.

The 95 percent confidence interval for β_1 is given by

$$-44.2 \pm 3.182 \, (3.5)$$

$$[-55.3, -33.1]$$

The Correlation Coefficient

Often in statistical analysis it is desirable to determine the strength of the relationship between the variables under study. The most widely used measure of this degree of association between Y and X is provided by r, the *coefficient of correlation*. The formula for r is

$$r = \frac{\sum xy}{\sqrt{\sum x^2 \, \sum y^2}}$$

The values of r lie in the interval $-1 \le r \le +1$ with a "large" value of r (either positive or negative) indicating a strong relationship between X and Y. A negative value of r indicates that high X values are associated with low Y values, or, low X values associated with high Y values. A positive r, on the other hand, indicates that high values of X are associated with high values of Y and low values of X are associated with low values of Y.

A further explanation of r may be seen by comparing it with $\hat{\beta}_1$, the slope of the regression line. In the formulas for r and $\hat{\beta}_1$, numerators are identical (the denominators for both will always be positive); therefore, r and $\hat{\beta}_1$ will have the same sign. When the slope of the line is negative, the correlation is also negative thus indicating a negative, or *inverse* relationship between Y and X. Similarly, a positive slope and a positive correlation indicate a *direct* relationship between variables. Further, if an *exact positive* relationship exists between Y and X (i.e., all points lie exactly on the regression line), then the value of r is $+1$. An *exact negative* relationship will yield an r of -1. When $\hat{\beta}_1 = 0$, $r = 0$ and hence no linear relationship between Y and X is indicated. (Caution: while r and $\hat{\beta}_1$ always have the same sign, the size of r is not indicative of the size of $\hat{\beta}_1$.)

As was the case with $\hat{\beta}_1$, the value r is the sample estimate of a true population correlation value denoted by ρ and is subject to sampling variation. It is of interest therefore to test the hypothesis that the true population correlation equals zero. A value of $\rho = 0$ indicates that there is no linear association between the variables under study. The test statistic for testing $H_0 : \rho = 0$ is

$$t = r \sqrt{\frac{n-2}{1-r^2}}, \qquad n-2 \text{ degrees of freedom}$$

A significant r indicates that the Y values are meaningfully related to the X values.

A simpler method for testing $\rho = 0$ is by comparing the value of r with values in Table D in the Appendix. If the absolute value of r exceeds the tabulated value, then r is said to be significant at the given α level.

Precautions in the Use of Correlation and Regression

In the interpretation of both the regression line and the correlation coefficient, there are several important precautions that must be considered.

The first of these is that the relationship between variables must be linear. A slope (β_1) or correlation coefficient (ρ) equal to zero does not imply that no relationship exists between the variables. It simply implies that there is no linear relationship between the variables. That is, the true functional form of the relationship may be quadratic, exponential, logarithmic, and so forth.

When the relationship between variables is not linear, the investigator may make transformations on one or both of the variables to "linearize" the data. A common example of this is the log-dose response curves used in drug studies. The relationship between dose level of the drug and response is often nonlinear, whereas the relationship between response and log dose is often linear. Thus, by transforming all dose units to log dose units, one may apply the principles of linear regression and correlation in the analysis of the data.

The second precaution that must be exercised in the interpretation of linear regression and correlation concerns the danger of making inferences beyond the range of actual observations upon which the analysis is based. Consider the example of dose-response relationships in pharmacology. The relationship between log dose and response may be described reasonably well by a straight line within the range of the doses applied, but as the range is extended to larger and larger dose levels, the same linear relationship may no longer hold. Another common danger is the extension of the regression line to $X = 0$. Often for very small X values the values of Y are decidedly nonlinear.

The third precaution that must be considered is that correlation does not necessarily mean causation. A significant correlation indicates that the two variables X and Y tend to be associated. Except for highly controlled studies in which all extraneous factors have been removed, it is impossible to determine which variable influences which, or even whether either of the variables is influencing the other directly. Often, a third variable may be affecting the relationship and "causing" both X and Y to vary together.

Summary of Regression and Correlation

1. Regression line: $\hat{Y} = \hat{\beta}_0 + \hat{\beta}_1 X$
2. Estimates of $\hat{\beta}_0$, $\hat{\beta}_1$:

$$\hat{\beta}_1 = \frac{\sum xy}{\sum x^2}$$

$$\hat{\beta}_0 = \bar{Y} - \hat{\beta}_1 \bar{X}$$

$$\sum x^2 = \sum(X - \bar{X})^2 = \sum X^2 - \frac{(\sum X)^2}{n}$$

$$\sum y^2 = \sum(Y - \bar{Y})^2 = \sum Y^2 - \frac{(\sum Y)^2}{n}$$

$$\sum xy = \sum(X - \bar{X})(Y - \bar{Y}) = \sum XY - \frac{(\sum X)(\sum Y)}{n}$$

Note: $\sum XY \neq (\sum X)(\sum Y)$.

3. Estimate of variation about regression line:

$$S_{y.x} = \sqrt{\frac{\sum y^2 - \hat{\beta}_1 \sum xy}{n - 2}}$$

4. Tests of hypothesis concerning $\beta_1 : (H_0 : \beta_1 = 0)$:

Large n: $z = \dfrac{\hat{\beta}_1 - \beta_1}{s_{y.x}/\sqrt{\sum x^2}} = \dfrac{\hat{\beta}_1}{s_{y.x}/\sqrt{\sum x^2}}$

Small n: $t = \dfrac{\hat{\beta}_1 - \beta_1}{s_{y.x}/\sqrt{\sum x^2}} = \dfrac{\hat{\beta}_1}{s_{y.x}/\sqrt{\sum x^2}},$ $df = n - 2$

5. Confidence intervals on β_1:

$$\hat{\beta}_1 \pm \theta_{\alpha/2} s_{y.x}/\sqrt{\sum x^2}, \qquad \theta = z \text{ or } t$$

6. Correlation coefficient:

$$r = \frac{\sum xy}{\sqrt{\sum x^2 \sum y^2}}$$

7. Test statistic for r:

$$t = r\sqrt{\frac{n - 2}{1 - r^2}}, \qquad df = n - 2$$

METHODOLOGY

An Example

A problem of great interest to college admissions committees is that of predicting a student's grade point average at the end of his freshmen year based on his score on an entrance exam taken prior to his first year.

The first step in obtaining such a prediction equation is to record both freshman GPA(Y) and entrance exam scores (X) for a random sample of n students selected from the population of interest (e.g., all freshman medical students in the United States). Using the sample data, a prediction equation is obtained by the method of least squares. This equation may then be used to predict the freshman GPA for any incoming student based on his entrance exam score.

Consider the data in Table 5.4 and the scatter diagram in Figure 5.5. The first step is to look at the scatter diagram. Notice how Y appears to increase as X increases. (What type of relationship does this indicate?)

Table 5.4. Entrance test scores and grade point average for 10 first-year students

Student	Test Score	GPA at the End of Freshman Year
1	24	1.5
2	61	3.5
3	30	1.7
4	48	2.7
5	60	3.4
6	32	1.6
7	19	1.2
8	22	1.3
9	41	2.2
10	46	2.7

We wish to find a regression equation of the form

$$\hat{Y} = \hat{\beta}_0 + \hat{\beta}_1 X$$

The necessary calculations are as follows:

$$\sum Y = 21.8 \quad \sum X = 383 \quad \sum Y^2 = 54.06 \quad \sum XY = 951.1$$

$$\bar{Y} = \frac{21.8}{10} = 2.18 \quad \bar{X} = \frac{383}{10} = 38.3 \quad \sum X^2 = 16767$$

$$\sum x^2 = \sum X^2 - \frac{(\sum X)^2}{n} = 16767 - \frac{(383)^2}{10} = 2098.1$$

$$\sum y^2 = \sum Y^2 - \frac{(\sum Y)^2}{n} = 54.06 - \frac{(21.8)^2}{10} = 6.536$$

$$\sum xy = \sum XY - \frac{(\sum X)(\sum Y)}{n} = 951.1 - \frac{(383)(21.8)}{10} = 116.16$$

Thus,

$$\hat{\beta}_1 = \frac{\sum xy}{\sum x^2} = \frac{116.16}{2098.1} = 0.0554$$

$$\hat{\beta}_0 = \bar{Y} - \hat{\beta}_1 \bar{X} = (2.18) - (0.0554)(38.3) = 0.0582$$

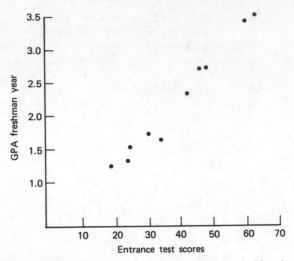

Figure 5.5. Scatter diagram of test scores and GPA at end of freshman year.

Therefore, the best fitting straight line relating entrance exam scores and freshman grade point average would be

$$\hat{Y} = 0.0582 + 0.0554X$$

Next, we wish to test whether $\hat{\beta}_1$ is significantly different from zero. That is, we wish to determine if a linear relationship exists between Y and X or if the value $\hat{\beta}_1 = 0.0554$ would have occurred by chance when, in reality, the true value of β_1 is zero. To perform this test of hypothesis we must first determine the variation around the regression line.

$$\text{SSE} = \sum y^2 - \hat{\beta}_1 \sum xy = 6.536 - (0.0554)(116.16)$$
$$= 0.1007$$

$$s_{y.x} = \sqrt{\frac{\text{SSE}}{n-2}} = \sqrt{\frac{0.1007}{8}} = 0.1122$$

Since n is small, we will use the t statistic to test the hypothesis of no linear relationship between variables.

STEP 1. $H_0:\beta_1 = 0$

STEP 2. $\hat{\beta}_1 = 0.0554$

STEP 3. $\quad t = \dfrac{\hat{\beta}_1 - 0}{s_{y.x}/\sqrt{\sum x^2}} = \dfrac{0.0554}{.1122/\sqrt{2098.1}} = \dfrac{0.0554}{.0024} = 23.1$

STEP 4. For $\alpha = 0.05$ and $df = n - 2 = 8$

$\qquad t_{\alpha/2} = \pm 2.306$

STEP 5. Since the calculated t lies in the rejection region we reject the null hypothesis of no linear relationship between variables. That is, we are 95 percent certain that $\hat{\beta}_1 = 0.0554$ is significantly different from zero and hence did not occur by random chance alone.

Since $\hat{\beta}_1$ is an estimate based on sample data, we may wish to determine the range in which we can reasonably expect the true population value β_1 to fall (with 95 percent confidence).

$$CI(\beta_1) = \hat{\beta}_1 \pm t_{\alpha/2} \frac{s_{y.x}}{\sqrt{\sum x^2}}$$

$$= 0.0554 \pm 2.306 \frac{.1122}{\sqrt{2098.1}}$$

$$= [0.050, 0.061]$$

Interpretation of regression line. Using the regression equation, we are able to predict Y for a given value of X. Suppose, for example, a student has an entrance exam score of 40; we would predict his freshman GPA to be

$$\hat{Y} = 0.0582 + 0.0554(40) = 2.27.$$

Correlation coefficient. The above regression line gives us no information about the strength of the relationship between Y and X. This information is provided by the correlation coefficient, r.

$$r = \frac{\sum xy}{\sqrt{\sum x^2 \sum y^2}} = \frac{116.16}{\sqrt{(2098.1)(6.536)}} = 0.99$$

By comparing the tabulated r from Table D in the Appendix with the calculated r, we see that a significantly strong linear relationship exists between Y and X.

The null hypothesis $H_0: \rho = 0$ may be tested by calculating the test statistic

$$t = r\sqrt{\frac{n-2}{1-r^2}}, \qquad df = n - 2$$

$$t = 0.99\sqrt{\frac{8}{1 - .99^2}}, \qquad df = 8$$

$$t = 19.8$$

For $\alpha = 0.05$ and $df = 8$ the critical t value is ± 2.306. We therefore reject the null hypothesis and conclude that a significant and useful predictive relationship exists between X and Y.

PROBLEMS

1. An experiment was conducted to study the effect of increasing the dosage of a certain barbiturate on sleeping time. Three readings were made at each of three dose levels.

Sleeping Time (Hours) (Y)	Dosage (μM/kg) (X)
4	3
6	3
5	3
9	10
8	10
7	10
13	15
11	15
9	15

$\sum Y = 72 \qquad\qquad \sum X = 84$

$\sum Y^2 = 642 \qquad\qquad \sum X^2 = 1002$

$\sum XY = 780$

a. Plot the scatter diagram.
b. Determine the regression line relating dosage (X) to sleeping time (Y).
c. Place a 95 percent confidence interval on β_1.
d. Test the hypothesis of no linear relationship between variables. (Use $\alpha = 0.05$.)
e. What is the predicted sleeping time for a dose of 12 μM/kg?

2. In a study of the relationship between amphetamine metabolism and

amphetamine psychosis, six chronic amphetamine users were given a psychosis intensity rating score. Plasma amphetamine levels (mg/ml) were also determined for these patients.

Patient #	Psychosis Intensity Rating	Plasma Amphetamine (mg/ml)
1	15	150
2	40	100
3	45	200
4	30	250
5	55	250
6	30	500

$$\sum X : 215 \qquad \sum Y : 1450$$
$$\sum X^2 : 8675 \qquad \sum Y^2 : 447500$$
$$\sum XY : 51500$$

a. Plot the scatter diagram.
b. Determine the correlation coefficient.
c. Is there a significant correlation between intensity rating and plasma amphetamine level? (Use $\alpha = 0.05$)

3. An investigator studying the effects of stress on blood pressure subjected nine monkeys to increasing levels of electric shock as they attempted to obtain food from a feeder. At the end of a 2-minute stress period blood pressures were recorded. (The nine monkeys had essentially identical initial blood pressure readings.)

Blood Pressure (Y)	125, 130, 120	150, 145, 160	175, 180, 180
Shock Intensity (X)	30	50	70

Intermediate Calculations

$$\sum X = 450 \qquad \sum Y = 1365$$
$$\sum X^2 = 24900 \qquad \sum Y^2 = 211475$$
$$(\sum X)^2 = 202500 \qquad (\sum Y)^2 = 1863225$$
$$\sum XY = 71450 \qquad (\sum X)(\sum Y) = 614250$$

a. Plot the scatter diagram.
b. Determine the regression line relating blood pressure to intensity of shock.
c. Place a 95 percent confidence interval on β_1.
d. Test the null hypothesis of no linear relationship between blood pressure and shock intensity (stress level). (Use $\alpha = 0.01$.)
e. For an intensity level of 60, what is the predicted blood pressure reading?

4. The following tabulation gives grade point average at the end of the first 2 years of basic sciences and scores on National Boards Part 1 for 12 medical students:

Student	GPR(X)	Score on National Board (Y)
1	3.35	620
2	2.37	445
3	3.13	445
4	3.10	560
5	1.94	295
6	3.00	570
7	2.85	415
8	1.96	430
9	2.98	560
10	2.55	515
11	2.23	430
12	1.95	435

$$\sum X = 31.41 \qquad \sum Y = 5720$$
$$\sum X^2 = 85.1323 \qquad \sum Y^2 = 2816050$$
$$\sum XY = 15357.55$$

a. Plot the scatter diagram.
b. Determine the correlation between GPR and National Board scores.
c. Is the correlation significant? (Use $\alpha = 0.05$.)
d. Determine the regression line relating GPR and National Board scores.
e. For a GPR of 3.0, what is the predicted National Board score?

5. In a study on the elimination of a certain drug in man, the following data were recorded:

Time in Hours(X)	Drug Concentration $(\mu g/ml)(Y)$
.5	.42
.5	.45
1	.35
1	.33
2	.25
2	.22
3	.20
3	.20
4	.15
4	.17

$$\sum X = 21 \qquad \sum Y = 2.74$$
$$\sum X^2 = 60.5 \qquad \sum Y^2 = 0.8526$$
$$\sum XY = 4.535$$

a. Plot the scatter diagram.
b. Determine the regression line relating time (X) to concentration of drug (Y).
c. Place a 99 percent confidence interval on β_1.
d. Test the null hypothesis of no relationship between variables. (Use $\alpha = 0.01$.)
e. What is the predicated drug concentration after 2 hours?

CHAPTER 6

EXPERIMENTAL DESIGN AND ANALYSIS OF VARIANCE

OVERVIEW

In Chapter 4 we considered the case of testing hypotheses concerning the means of one (simple z or t) or two (pooled or paired t) populations. The purpose of this chapter is to extend these techniques to cases involving more than two populations. The appropriate statistical method for accomplishing this is known as the analysis of variance (ANOVA). It involves the comparison of the amount of variance among observed mean values that can be attributed to actual population differences to the variation attributable to random chance. As we will see, the analysis of variance is an arithmetic procedure for partitioning the total variation exhibited by a set of data into recognizable sources of variation such as treatment differences, classification differences, and residual variation (experimental error).

OBJECTIVES

Given a narrative description of an experimental situation, the objectives for this unit are for you to be able:

- To select and justify the choice of an appropriate statistical experimental design
- To carry out the correct method of analysis of the chosen design
- To interpret the results of the analysis

FOUNDATIONS

Experimental Design

Variation in results is the basic characteristic of all experiments. The source of this variation is twofold. First, there is the inherent variability among the objects being measured. The objects being measured are called *experimental units*. Second, there is the variation associated with the lack of uniformity in the conditions and execution of the experiment. We attempt to reduce variation from these two sources by obtaining homogeneous experimental units, refining the experimental technique, and conducting the experiment by the most exact methods.

In previous chapters estimates of experimental error were used in the form of standard deviations or standard errors. Thus, more than one experimental unit must be observed under similar conditions in order to estimate experimental error. If more than one experimental unit is observed under similar conditions, the experiment is said to be replicated. The precision of the estimate of experimental error as well as the stability of the estimates of means is increased with increased replication. Recall that the standard error of a mean is given by $s/\sqrt{n}$, which decreases as n increases.

In addition to providing for adequate numbers of subjects, there are other basic principles that are fundamental to the design and execution of all experiments. A well-conceived experiment has five parts:

1. Statement of the experimental objectives
2. Specification of the hypotheses to be tested
3. Definition of experimental units and treatments
4. The experimental design
5. Method of analysis

If all these are well thought out, the investigator can make valid statistical inferences from his experimental data. We shall now discuss these general considerations relevant to the design and analysis of experiments.

Statement of objectives.   This statement often takes the form of the means to be estimated, hypotheses to be tested, or questions to be answered. All statements should be lucid and specific. The most common fault of an experimental design is vagueness and excessive expectation of the results. A careful statement of the research objectives also includes the identification of the population of inference. This is particularly important in evaluating the adequacy of randomization and sampling schemes.

Specification of the null hypothesis.   The null hypothesis, as indicated earlier, is generally a simple statement of no difference between two or more population means. These may be distinct populations with suspected natural or environmental differences or samples from a single population with each sample treated differently. From the data, estimates are made of natural or environmental differences or of treatment effects. Then, under the null hypothesis, the probability associated with the appropriate test statistic is computed from the data. If this probability is small, the experimenter concludes that the null hypothesis must have been erroneous and he accepts the hypothesis that there is indeed a difference among the populations being studied.

Experimental units and treatments.   The experimental unit is the object to which a treatment or procedure is applied or a member of a specified population. The purpose of the experiment is to measure differences among specified populations or compare the effects of the treatments or procedures.

For the purposes of statistical discourse, any difference in the way samples are classified is called a *treatment*. This simplifies discussion and helps unify the methodology.

The specification of the treatments in a medical experiment may raise difficult questions about the conditions under which observations are to be made. A concise statement of the experimental objectives is fundamental to the resolution of such questions. Often also there is a question of whether a control group is needed. In general, a *control* is a treatment that has no particular experimental interest but which is needed in order to reveal whether another treatment is effective. Some considerations about the use of a control are the following:

1. The effectiveness of a standard treatment could have been so consistently demonstrated by previous experimentation that there is no need for a control. All that is in order is to determine the effect of the experimental material.
2. The effect of the standard treatment could be well known under certain conditions, but it could be possible that its effect would be modified by the new experimental conditions. In such a case, a control group would serve primarily to describe the consistency of the experiment with previously reported results.
3. There may be no information about the efficacy of any treatment. In such cases, a control should be included. There may be a case for greater replication of this treatment than of the other treatments in order to obtain useful baseline data.

In the design of experiments special attention is paid to randomization procedures. The function of randomization is to insure that each treatment has an "equal chance" to be assigned to any particular subject or procedure. Thus, no treatment receives a preferential assignment of subjects. It has been shown that a systematic assignment of experimental units to treatment groups often results in either underestimation or overestimation of experimental error. An example of systematic assignment to treatment groups would be asking cancer patients whether they would undergo surgical treatment and then assigning all who said no to a medical regimen for comparison. A better way would be to divide randomly the group that volunteered for surgery into surgical and medical treatment groups for comparison of the techniques.

It should also be kept in mind that haphazardness and randomness are not equivalent. One might feel that reaching into a cage of mice and taking the first ones to come to hand as a control group would provide the desired randomness. However, this might be selecting only those mice that are sick or otherwise lethargic for some unknown reason.

As one might suppose, there is a large array of experimental designs from which to choose. Each of these designs should cover such matters as the experimental treatments or classifications, the size of the experiment, the experimental method, identification of the proper control or comparison groups, the method of randomization, and the steps made to insure that all groups are handled similarly except for the treatment itself.

The choice of a design is based on recognition of the sources of variation inherent in the system being studied. Only two of the more common experimental designs will be discussed in this chapter. They are the completely random design and the randomized complete block design.

Completely Random Design

The *completely random design* (CRD) may be considered an extension of the pooled *t* situation discussed in Chapter 4. In this design, experimental units are simply chosen at random from the population to which inferences are to be made. The total sample is randomly divided into groups and the different treatments or conditions under study are then applied to the groups, one treatment or condition to a group. If the treatments differ from each other then the various treatment groups will have different mean values at the end of the experiment. Examples of a CRD are given below.

Example 1. A nurse practitioner was interested in evaluating three different back-rub techniques on patient blood pressure. Fifteen patients were selected for the experiment and each was randomly assigned to one of the three treatments (back-rub techniques). Following the application of the treatment, systolic blood pressure was recorded.

	Treatments	
Procedure 1	*Procedure 2*	*Procedure 3*
130	140	125
145	140	118
120	155	120
130	120	140
125	130	125

Example 2. In an experiment to compare the effect of four different diets on reduction of serum cholesterol levels, men were assigned at random each to one of the four treatment (diet) groups. No effort was made to have equal numbers in each group. At the end of 3 months, serum cholesterol determinations were made on each of the participants.

	Treatments		
Diet 1	Diet 2	Diet 3	Diet 4
200	240	180	260
260	245	220	260
220	260	245	280
235	235		230
240			245

Example 3. Three instruments in a clinical chemistry laboratory were evaluated in terms of their ability to produce similar results. A blood sample was divided into 12 equal portions and each portion randomly assigned to an instrument. The following quadruplicate blood sugar determinations (in mg/100 ml) were made.

	Treatments	
Instrument 1	Instrument 2	Instrument 3
118	108	115
118	112	112
120	115	110
110	125	120

Randomized Complete Block Design

For some experiments, it may be possible to "block" experimental units into homogeneous groups. A block can be considered an extension of the concept of a paired experiment as discussed in Chapter 4. In this case, the experimental units are subdivided into homogeneous groups called blocks, and the treatments are then randomly assigned to members of the block. Each block receives every treatment and each treatment appears in every block. This is called the randomized complete block design (RCB).

The randomized complete block is used when the investigator feels that the response of the experimental unit may be affected by a factor or factors other than the treatment under study. For example, patients in a drug study may be in different age groups. It is likely that response to drug would be affected by age of the patient. By blocking the experimental units (patients) into homogeneous age groups, the variation attributable to the extraneous factor (age) can be removed. Examples of a RCB are given below.

Example 1. In a study of weight gain associated with three supplemental diets, it was felt that initial weight of the participants would be reflected in the final response to the diet. The patients were divided according to weight at the outset of the experiment. At the end of a 3-month period the weight gain of each participant was recorded.

Block Initial Weight	Diet 1	Treatments Diet 2	Diet 3
100–120	10	12	8
120–140	8	10	10
140–160	5	12	10
160–180	10	5	5

Note: Within each block, the participants were randomly assigned to the treatments (diet).

Example 2. An investigator wished to compare the elapsed time of three different methods of collecting information on patients in an emergency room. Three different hospital emergency rooms were utilized in the study. Response is time (in minutes) to collect the information.

Block Hospital	Method 1	Treatments Method 2	Method 3
1	30	25	20
2	20	15	20
3	40	35	25

Example 3. In evaluating three instruments in a laboratory, three different bottles of standard were used. It was felt that the differences in the standard might affect the readings of the instruments.

Block Standard	Instrument 1	Treatments Instrument 2	Instrument 3
1	124	111	117
2	116	101	115
3	120	115	122

Aliquots of standard within each block were randomly assigned to the instruments.

Analysis of Variance

The final specification for a well-conceived experiment is the method of analysis. Each experiment will have its own appropriate analysis. For the completely random design and the randomized complete block design, the general method is the *analysis of variance*. The process of using the ANOVA (analysis of variance) is best learned by studying examples.

METHODOLOGY

Completely Randomized Design

Formally stated, in a completely randomized design there are k treatments, each of which is assigned at random to a group of experimental units. The null hypothesis is whether the treatment means are all equal. Symbolically, $H_0: \mu_1 = \mu_2 = \cdots = \mu_k$ which is tested to see whether the treatment groups are really subsamples from the same population (H_0 true) or whether they are samples from different populations (H_0 false). The pooled t-test described in Chapter 4 is one case of the completely randomized design ($k = 2$).

In a completely randomized design each experimental unit has an equal and independent chance of receiving any one of the treatments. The basic assumption underlying this design is that the observed values in any one group represent a random sample of all possible values of all experimental units under that particular treatment. Further, we assume that the responses are normally distributed about the treatment mean and that the variation among observations treated alike is identical for all treatments.

Table 6.1 gives cholesterol levels (in mg/100 ml) for normal men on four experimental diets. Since in this example the criterion of classification is diet, it is of interest to test whether the means of all four diets are significantly different, that is, do these diets have an effect on cholesterol level in normal men? Each participant was assigned at random to one of the diets; therefore, the appropriate statistical model is that of the completely randomized design. Calculations that will be useful for computing the ANOVA are also shown in the table.

The null hypothesis to be tested is $H_0: \mu_A = \mu_B = \mu_C = \mu_D$ against the alternative that at least one pair of the means is not equal. In words, we

Table 6.1. Cholesterol level (in mg/100 ml) for normal men on four different diets

		Diet				
		A	B	C	D	
		200	240	180	260	
		260	245	220	260	
		220	260	245	280	
		235	235		230	
		240			245	
Total	(T_i)	1155	980	645	1275	4055
	n_i	5	4	3	5	
	$\bar{Y}_i$	231	245	215	255	

wish to determine if the true population mean cholesterol levels for the four diet groups are different. If the population means are different then it is assumed that diet has an effect on serum cholesterol level in normal males.

If the null hypothesis is true, the variation between treatment groups should not be significantly different from that expected from the variation within groups. We may doubt the validity of H_0 if it can be shown that the between group variation is significantly larger than what is expected based on the within group variation. (If H_0 is true and the treatment groups are from the same population, then the within and the between variation can both be used to estimate the population variance σ^2.)

Calculations from analysis of variance techniques are customarily displayed in an ANOVA table as depicted in Table 6.2. Definitions and computing formulas for the terms shown in Table 6.2 are discussed below.

Table 6.2. ANOVA for the completely randomized design

Source of Variation	Degrees of Freedom	Sum of Squares	Mean Squares	F
Among treatments	$k-1$	SST	$MST = SST/(k-1)$	MST/MSE
Within treatments	$N-k$	SSE	$MSE = SSE/(N-k)$	
Total	$N-1$	SS		

Total sum of squares (SS). As stated previously, the rationale behind the analysis of variance technique is to partition the total variation in a set of data into recognizable sources such as treatment differences and experimental error. For ease of discussion sum of squares rather than actual variances are partitioned.

The total sum of squares is the total of the squared deviations of the observations from the overall mean of the data. It is simply the numerator in the familiar formula for calculating the variance of all the observations considered as a single group. Symbolically,

$$SS_{Total} = \sum_{all} Y^2 - \frac{\left(\sum_{all} Y \right)^2}{N}$$

where

$$N = n_1 + n_2 + n_3 + \cdots + n_k, \; k = \text{number of treatments}$$

For convenience of calculations, the term

$$\frac{\left(\sum_{\text{all}} Y\right)^2}{N}$$

is given a special name. It is called the *correction factor* and is used in several calculations.

For the data in Table 6.1,

$$CF = \frac{\left(\sum_{\text{all}} Y\right)^2}{N} = \frac{(4055)^2}{17} = 967,236.7647$$

and

$$SS_{\text{Total}} = \sum_{\text{all}} Y^2 - CF = 976,625 - 967,236.7647$$

$$= 9388.2353$$

Within groups sum of squares (SSE). Since the within treatments variation is the variation associated with observations treated alike, it is the variation associated with experimental or random error. As would be expected, to obtain a numerical value for this within group variation, we obtain a measure of the variation within each treatment group and combine these variance contributions to form a pooled estimate.

Recall from the pooled t situation that the pooled variance estimate was

$$s_p^2 = \frac{(n_1 - 1)s_1^2 + (n_2 - 1)s_2^2}{n_1 + n_2 - 2}$$

where s_1^2 and s_2^2 were the variances of the two samples. For the k sample case the logical extension to obtain the pooled estimate of within group variation is

$$s_w^2 = \frac{(n_1 - 1)s_1^2 + (n_2 - 1)s_2^2 + \cdots + (n_k - 1)s_k^2}{n_1 + n_2 + \cdots + n_k - k}$$

This may be rewritten

$$s_w^2 = \frac{SSE}{N - k} = MSE \text{ (mean square error)}$$

since $n_1 + n_2 + \cdots + n_k = N$, the total number of observations. This formula is generally not used for computations unless it is desired to have available the standard error for each treatment group. The computational formula for SSE is given by

$$SS_{Within} = SSE = \sum_{all} Y^2 - \sum_{i=1}^{k} \frac{(T_i)^2}{n_i}$$

For our example

$$SSE = 976625 - \left[\frac{(1155)^2}{5} + \frac{(980)^2}{4} + \frac{(645)^2}{3} + \frac{(1275)^2}{5} \right]$$

$$= 5920$$

Among groups sum of squares (SST).   The final source of variation to be calculated is the among treatments variation (the failure of the k treatment means to be alike). The computational formula is given by

$$SS_{Among} = SST = \sum_{i=1}^{k} \frac{(T_i)^2}{n_i} - CF$$

For our example,

$$SST = \frac{(1155)^2}{5} + \frac{(980)^2}{4} + \frac{(645)^2}{3} + \frac{(1275)^2}{5} - 967,236.7647$$

$$= 3468.2353$$

A final calculational short-cut may be developed by utilizing the relationship

$$SS_{Total} = SS_{Within} + SS_{Among}$$

$$= SSE + SST$$

In practice, SSE is rarely computed directly. Rather it is obtained by subtraction, that is,

$$SSE = SS_{Total} - SST$$

The general procedure for computing the mean square column for the ANOVA is to compute first the sum of squares and enter in the ANOVA table; then compute the degrees of freedom and enter in the table; and,

Table 6.3. ANOVA of cholesterol levels

Source	Degrees of Freedom	Sum of Squares	Mean Squares	F
Between	3	3468.2353	1156.08	2.54
Within	13	5920	455.38	
Total	16	9388.2353		

finally to compute the mean square by dividing the degrees of freedom into the sum of squares. (See Table 6.3.)

For our example,

$$MST = \frac{SST}{k-1} = \frac{3468.2353}{3} = 1156.08$$

and

$$MSE = \frac{SSE}{N-k} = \frac{5920}{13} = 455.38$$

The test of the significance of differences among means is accomplished by computing the ratio of the estimate of σ^2 based on between variation (MST) to the estimate based on within variation (MSE). This ratio is called an F *statistic*. The larger this ratio, the greater the difference between the two values and the less likely the null hypothesis is true. Therefore, for large F, we reject H_0 and conclude that the means of the treatment groups are significantly different; the groups are not drawn from the same population. Symbolically,

$$F = \frac{MST}{MSE}$$

If the null hypothesis is true and $\mu_A = \mu_B = \mu_C = \mu_D$, then MST and MSE are both estimates of the common variance σ^2 of the population. Under this assumption the F ratio MST/MSE should be close to one. Actually, the calculated value is

$$F = \frac{MST}{MSE} = \frac{1156.08}{455.38} = 2.54$$

To determine if the calculated F value is large enough to warrant rejection of H_0 we use Table E in the Appendix to locate the tabulated critical value, F_α. The degrees of freedom associated with F are γ_1 and γ_2 where

$$\gamma_1 = df \text{ associated with numerator (MST)}$$

$$\gamma_2 = df \text{ associated with denominator (MSE)}$$

The degrees of freedom associated with the numerator (γ_1) determines the appropriate column in the table; the denominator degrees of freedom (γ_2) determines the appropriate row. Thus, in this example we use the column numbered 3 and row numbered 13. The critical values for $\alpha = 0.05$ and $\alpha = 0.01$ are 3.41 and 5.74, respectively. Since our observed $F = 2.54$ is less than both the $\alpha = 0.05$ and $\alpha = 0.01$ critical values, there is insufficient evidence to show that the population means are not equal. It cannot be said that the diets studied have an effect on cholesterol level in normal males.

Randomized Complete Block Design

Three large hospitals agreed to a cooperative study of laboratory costs by hospital service. In each hospital a patient on each of the various service wards was selected at random and lab fees were monitored. The data are shown in Table 6.4. Thus, the hospitals are considered "blocks" on the basis

Table 6.4. Patient laboratory costs (dollars) by hospital and type of service

Blocks (Hospital)	Surgery	Medicine	Pediatrics	Ob-Gyn
Hospital 1	44	59	60	41
Hospital 2	33	19	49	71
Hospital 3	44	40	45	31

Treatments (Service) spans Surgery, Medicine, Pediatrics, Ob-Gyn.

that laboratory fees within a hospital should be uniform across types of service. That is, a hospital with low lab fees should be low for all patients regardless of whether they are surgical, medical, and so forth. The types of service are the "treatments" being studied.

The rationale for computing the ANOVA for the randomized complete block design follows that of the completely randomized design with the simple addition of calculating the sum of squares for blocks. The symbolic form for the ANOVA is shown in Table 6.5.

Table 6.5. ANOVA for the randomized complete block

Source of Variation	Degrees of Freedom	Sum of Squares	Mean Squares	F
Among blocks	$b-1$	SSB		
Among treatments	$k-1$	SST	$MST = SST/(k-1)$	MST/MSE
Error	$(b-1)(k-1)$	SSE	$MSE = SSE/(b-1)(k-1)$	
Total	$bk-1$	SS		

The procedure for computing the needed sums of squares is summarized as follows:

STEP 1. Compute the correction factor (CF) as for the completely randomized design.

STEP 2. Compute the block sum of squares (SSB) by summing the squares of block totals, dividing this sum by the number of treatments, and subtracting the correction factor.

STEP 3. Compute the treatment sum of squares, SST (as before), by summing the squares of the treatment totals, dividing by the number of blocks, and subtracting the correction factor.

STEP 4. Compute the error sum of squares (SSE) by subtracting the block sum of squares and the treatment sum of squares from the total sum of squares (SSE = SS − SSB − SST).

STEP 5. Fill in and complete the ANOVA as in Table 6.5.

The individual observations and the necessary totals are shown in Table 6.6. Note: The B_j are block totals and the T_i are treatment totals. There are b blocks and k treatments.

Table 6.6. Preliminary calculations for the ANOVA of laboratory costs (Table 6.4)

Blocks (Hospitals)	Treatments (Service)				
	Surgery	Medicine	Pediatrics	Ob-Gyn	Totals (B_j)
1	44	59	60	41	204
2	33	19	49	71	172
3	44	40	45	31	160
Totals (T_i)	121	118	154	143	536

1. Correction factor

$$CF = \left(\sum_{\text{all}} Y\right)^2 \bigg/ N = (536)^2/12$$

$$= 23941.3$$

where N = total number of observations = bk

2. Total sum of squares

$$SS_{\text{Total}} = \sum_{\text{all}} Y^2 - CF$$

$$= (44)^2 + (59)^2 + \cdots + (31)^2 - CF$$

$$= 26112 - 23941.3$$

$$= 2170.7$$

3. Sum of squares for treatments

$$SST = \sum_{i=1}^{k} \frac{(T_i)^2}{n_i} - CF$$

$$= \frac{(121)^2}{3} + \frac{(118)^2}{3} + \frac{(154)^2}{3} + \frac{(143)^2}{3} - 23941.3$$

$$= 302.0$$

4. Sum of squares for blocks

$$SSB = \sum_{j=1}^{b} \frac{(B_j)^2}{n_j} - CF$$

$$= \frac{(204)^2}{4} + \frac{(172)^2}{4} + \frac{(160)^2}{4} - 23941.3$$

$$= 258.7$$

5. Sum of squares for error

$$SSE = SS_{\text{Total}} - SST - SSB$$

$$= 2170.7 - 302.0 - 258.7$$

$$= 1610.0$$

The ANOVA can now be completed (Table 6.7). The tabulated F value for three (MST) and six (MSE) degrees of freedom is 4.76. Since the calculated value does not exceed the table value, it is concluded that there is insufficient evidence to show that patient laboratory costs differ among the various types of hospital services studied.

Table 6.7. ANOVA of laboratory cost

Source of Variation	Degrees of Freedom	Sum of Squares	Mean Squares	F
Among hospitals	2	258.7		
Among services	3	302.0	100.7	0.375
Error	6	1610.0	268.3	
Total	11	2170.7		

Interpretation of the ANOVA

The F-test for significance of differences among three or more mean values is a cumulative test in the sense that it takes into account the differences among all pairs of means but does not single out any particular pair of mean values as being different. All a significant F-test indicates is that at least some of the means differ among themselves. That is, at least one $\mu_i \neq \mu_j$, so the null hypothesis is rejected.

Usually, in addition to testing the global hypothesis that all population means are equal, we are interested in testing hypotheses about specific mean differences. That is, we wish to determine just which pairs of means are different.

Testing Hypotheses About Specific Differences Following the Analysis of Variance

One method that immediately comes to mind for testing differences between individual pairs of means is performing t-tests on all possible pairs of means. When, for example, a study consists of three treatments, we may test the hypotheses $\mu_1 = \mu_2$, $\mu_1 = \mu_3$, $\mu_2 = \mu_3$ by carrying out the two-sample t-test procedure as described in Chapter 4 for each hypothesis. There is, however, a pitfall associated with this "multiple t-test" procedure. When each individual comparison is tested at a specified level of significance, α, the corresponding level of significance for the set of C comparisons is at most $C\alpha$, where $C\alpha$ is called the *error rate per experiment*. Thus, if we do five comparisons, each at $\alpha = .05$, the per experiment error rate could be as much as $5 \times .05 = .25$, which far exceeds the overall 5 percent rate we usually want. To avoid this we make the individual comparisons at a smaller significance level as described below.

The per experiment error rate may be conceptualized as the average number of comparisons that would be falsely declared significant per experiment if, theoretically, the experiment was repeated many times. Here, the term experiment refers not only to the physical conduct of the study but also to the repeated testing of the set of comparisons defined by the investigator to be of interest. If, as usual, we want our per experiment error rate to be α, then we should test each of the C individual comparisons at the significance level α/C. That is, we would compare the computed t statistic to the tabulated t value of $t_{\alpha/C}$ for a one-tailed test or $t_{(\alpha/C)/2}$ for a two-tailed test. This procedure, called the *Bonferroni procedure*, ensures that the error rate for *all* the comparisons of interest considered as a group will be at most α per experiment.

Example. A study was carried out to compare four different drugs with respect to changes in blood pressure. Summary statistics and the ANOVA table are given below (Table 6.8).

		Drug		
	A	B	C	D
$\bar{Y}$:	9.0	13.0	11.0	15.0
n_i:	4	4	4	4
s:	1.826	1.826	1.826	1.826

Table 6.8. ANOVA

Source	df	SS	MS	F
Among treatments	3	80	26.7	8.1
Within treatments	12	40	3.3	
Total	15	120		

$F_{3,12,.05} = 3.49$

Suppose the investigator has defined the following five hypotheses to be of interest.

$$\mu_A = \mu_B$$

$$\mu_A = \mu_C$$

$$\mu_A = \mu_D$$

$$\mu_B = \mu_C$$

$$\mu_C = \mu_D$$

These individual hypotheses may be tested using the t-statistic

$$t = \frac{\bar{Y}_i - \bar{Y}_j}{\sqrt{MSE\left(\dfrac{1}{n_i} + \dfrac{1}{n_j}\right)}}$$

where MSE is the pooled estimate of within group variation obtained from the ANOVA table. The degrees of freedom for this test is the degrees of freedom for the within treatments sums of squares in the ANOVA table. Returning to the example, the five t statistics are summarized below.

Hypothesis	t statistic		
$\mu_A = \mu_B$	$\dfrac{\bar{Y}_A - \bar{Y}_B}{\sqrt{MSE\left(\dfrac{1}{n_A} + \dfrac{1}{n_B}\right)}}$	$= \dfrac{9.0 - 13.0}{\sqrt{3.3\left(\dfrac{1}{4} + \dfrac{1}{4}\right)}}$	$= -3.11$
$\mu_A = \mu_C$	$\dfrac{\bar{Y}_A - \bar{Y}_C}{\sqrt{MSE\left(\dfrac{1}{n_A} + \dfrac{1}{n_C}\right)}}$	$= \dfrac{9.0 - 11.0}{\sqrt{3.3\left(\dfrac{1}{4} + \dfrac{1}{4}\right)}}$	$= -1.56$
$\mu_A = \mu_D$	$\dfrac{\bar{Y}_A - \bar{Y}_D}{\sqrt{MSE\left(\dfrac{1}{n_A} + \dfrac{1}{n_D}\right)}}$	$= \dfrac{9.0 - 15.0}{\sqrt{3.3\left(\dfrac{1}{4} + \dfrac{1}{4}\right)}}$	$= -4.67$
$\mu_B = \mu_C$	$\dfrac{\bar{Y}_B - \bar{Y}_C}{\sqrt{MSE\left(\dfrac{1}{n_B} + \dfrac{1}{n_C}\right)}}$	$= \dfrac{13.0 - 11.0}{\sqrt{3.3\left(\dfrac{1}{4} + \dfrac{1}{4}\right)}}$	$= 1.56$
$\mu_C = \mu_D$	$\dfrac{\bar{Y}_C - \bar{Y}_D}{\sqrt{MSE\left(\dfrac{1}{n_C} + \dfrac{1}{n_D}\right)}}$	$= \dfrac{11.0 - 15.0}{\sqrt{3.3\left(\dfrac{1}{4} + \dfrac{1}{4}\right)}}$	$= -3.11$

To ensure that the overall error rate per experiment does not exceed $\alpha = .05$, we compare each calculated t statistic with the tabulated t value:

$$t_{.05/5} = t_{.01} = 2.681 \quad \text{(12 degrees of freedom)}$$

for a one-tailed test or

$$t_{(.05/5)/2} = t_{.005} = 3.0545$$

for a two-tailed test. Assuming that we are interested in the two-tailed alternatives $\mu_i \neq \mu_j$, we may conclude simultaneously that drug A is different from drug B and drug D, and drug C is different from drug D.

Suppose, instead of testing only the five hypotheses we have just illustrated, we are interested in testing all six possible pairwise hypotheses:

$$\mu_A = \mu_B$$

$$\mu_A = \mu_C$$

$$\mu_A = \mu_D$$

$$\mu_B = \mu_C$$

$$\mu_B = \mu_D$$

$$\mu_C = \mu_D$$

For the two-tailed alternative $\mu_i \neq \mu_j$, we compare the calculated t statistics with the tabulated value $t_{(\alpha/6)/2} = t_{(.05/6)/2} = t_{.004}$. This procedure often requires tabulated t values not found in standard t tables. One may use linear interpolation to obtain values from standard t tables, or an approximate value for t that cuts off the upper $\alpha/2$ proportion with v degrees of freedom can be determined from the standardized normal distribution by

$$t_{\alpha/2,v} = z + \frac{z^3 + z}{4(v - 2)}$$

where z is the corresponding value in a normal distribution.* Using this approximation,

$$t_{.004,12} = z_{.496} + \frac{z_{.496}^3 + z_{.496}}{4(10)}$$

$$= 2.65 + \frac{(2.65)^3 + 2.65}{40}$$

$$= 3.18$$

The six calculated t statistics appropriate for each of the six possible pairwise hypotheses are compared with the approximate table value of 3.18 to determine which pairwise differences are significant.

---

* Kirk, Roger E., *Experimental Design: Procedures for the Behavioral Sciences*. Belmont, California: Wadsworth Publishing Co., Inc., 1968.

When the number of comparisons is large, the Bonferroni procedure is very conservative. That is, as the number of comparisons increases, the tabulated value of t against which the calculated t statistics are compared increases, and it becomes increasingly difficult to declare a difference in means to be significant. Often it is more difficult than is really necessary. Under these circumstances, other multiple-comparison procedures may be preferred. These procedures are discussed in most intermediate-level statistics textbooks.

Summary of ANOVA
for the CRD and the RCB

The steps for computing the ANOVA for the CRD and RCB are:

STEP 1. Compute the square of the total of all observations and divide by the total number of observations. This is called the correction factor.

STEP 2. Compute the total sum of squares (SS) by summing the squares of all of the individual observations and subtracting the correction factor.

STEP 3. Compute the treatment sum of squares (SST) by summing the squares of treatment totals, each square divided by its own number of observations, and subtracting the correction factor.

STEP 4. For RCB compute the block sum of squares (SSB) by summing the squares of block totals, each square divided by its own number of observations, and subtracting the correction factor.

STEP 5. Compute the error sum of squares:
for CRD:

$$SSE = SS_{Total} - SST$$

for RCB:

$$SSE = SS_{Total} - SST - SSB$$

STEP 6. Compute degrees of freedom.

STEP 7. Compute mean squares by dividing each sum of squares by its own degrees of freedom.

STEP 8.   Calculate F statistic:

$$F = \frac{\text{MST}}{\text{MSE}}$$

STEP 9.   Determine rejection region from Table E using γ_1 (numerator) and γ_2 (denominator) degrees of freedom.

STEP 10.   If calculated F is greater than tabulated value from Table E, reject null hypothesis of equal population means.

PROBLEMS

1. Sixteen hospital patients were randomly assigned in groups of four to four treatment groups. Changes in blood pressure following treatment are shown below.

	Drug		
A	B	C	D
10	12	9	17
8	14	13	14
7	11	10	13
11	15	12	16

a. What is the experimental design?
b. List the means and standard errors for each drug.
c. Construct the ANOVA using the method of computing the error sum of squares directly. (Use $\alpha = 0.05$.)
d. Compute the total sum of squares directly and get the error sum of squares by subtraction. Compare to part c.
e. Interpret the ANOVA.

2. In a quality control study of clinical chemistry laboratories, samples of four different enzymes were sent to three commercial labs to test whether the labs were getting similar results.

	Enzyme			
Laboratory	1	2	3	4
A	4.2	6.0	3.9	8.3
B	3.9	7.3	4.0	7.2
C	5.2	6.5	3.2	6.9

a. What is the experimental design?
b. Test the hypothesis that the laboratories give similar results. (Assume that the experimental error is the same for each enzyme.) (Use $\alpha = 0.05$.) Interpret the results.

3. Given the following data:

Treatment Group

1	2	3
4	3	12
7	5	8
6	2	9
3		11
2		

a. What is the experimental design?
b. Test the hypothesis that the treatments have equal effects. (Use $\alpha = 0.01$.) Interpret the results.

4. Four treatments were compared in a randomized complete block as shown below:

	Treatment			
Block	1	2	3	4
1	4.0	3.1	4.4	5.9
2	6.6	6.4	3.3	1.9
3	4.9	7.1	4.0	4.0
4	7.3	6.7	6.8	3.1

Compute the ANOVA and test the hypothesis of equal treatment means at $\alpha = 0.01$. Interpret the results.

5. A double blind clinical study was conducted to compare the analgesic effect of Motrin (ibuprofen) (400 mg), codeine (60 mg), codeine (30 mg), and placebo. Twenty participants were randomly assigned in groups of 5 to the four treatment groups. Two hours after the administration of the treatment, patients were asked to rate degree of pain relief on a scale from 0 to 100. The following data were obtained

Motrin, 400 mg	Codeine, 60 mg	Codeine, 30 mg	Placebo
82	80	77	65
89	70	69	75
77	72	67	67
72	90	65	55
92	68	57	63

a. Do these data provide sufficient evidence at the .05 level of significance to indicate a difference in perceived pain relief among the four treatments?
b. If so, which pairs of treatments are different? (Use a per experiment error rate of $\alpha = .05$.)

6. In a study to evaluate hypoglycemic effectiveness, each of five maturity-onset diabetics was given a treatment that consisted solely of dieting, a treatment that consisted of chlorpropamide (100 mg/day), and one that consisted of chlorpropamide (250 mg/day). The order of administration of each treatment was assigned at random. At the end of a specified time period, Hb A_{1c}(percentage) levels were determined and the following data were obtained.

Patient	Diet Alone	Chlorpropamide (100 mg/day)	Chlorpropamide (250 mg/day)
1	8	5	5
2	7	6	5
3	9	8	7
4	7	5	5
5	8	6	7

a. Do these data provide sufficient evidence to indicate a difference in Hb A_{1c}(percentage) among the three different treatments? (Use $\alpha = .01$.)

b. If so, which treatments are different? (Use a per experiment error rate of $\alpha = .01$.)

7. A study was conducted to compare the effects of three different drugs on the alleviation of anxious depression in the nonpsychotic patient. Twelve nonpsychotic patients, all suffering from moderate to severe depression and anxiety, were assigned at random to the three treatment groups. A combined anxiety and depression score as determined by the MMPI and Taylor Manifest Anxiety Scale was recorded for each participant at the end of one week of therapy.

Drug A	Drug B	Drug C
25	20	25
15	16	15
20	18	20
14	25	20

a. Do the data suggest that there is a difference in anxiety/depression scores for the three drugs? (Use $\alpha = .05$.)

b. If so, which drugs are different? (Use a per experiment error rate of $\alpha = .05$.)

8. In a study similar to that described in problem 7, patients were grouped according to initial level of severity of anxiety and depression. Patients in each severity level group were randomly assigned to the three treatments (drugs). At the end of the experimental period, anxiety/depression scores were recorded.

Initial Severity Level	Drug A	Drug B	Drug C
1	35	30	25
2	40	25	20
3	25	25	20
4	30	25	25

a. Do the data suggest that there is a difference in anxiety/depression scores for the three drugs? (Use $\alpha = .05$.)

b. If so, which drugs are different? (Use a per experiment error rate of $\alpha = .05$.)

CHAPTER 7

ENUMERATION DATA

OVERVIEW

In previous chapters the z, t, and F statistics have been applied to measurement data that were either continuous or approximately so. In this chapter techniques are presented for dealing with discrete data arising from experimental or sampling situations where observations are simple counts.

Many experiments in medicine yield *enumerative* or count data (i.e., data consisting of the number of individuals in a sample falling into various categories). Examples of categories of classification are sex, age, race, financial bracket, improved or not improved, and so forth.

Enumeration data are generally qualitative rather than quantitative. The data consist of the counts, or proportions, of individuals in each classification. Statistical comparisons and inferences are made in terms of these proportions.

The technique for testing hypotheses concerning enumeration data is to compute what is known as a χ^2 (chi-square) statistic and compare it to tabulated critical values just as was done for the z, t, and F. The rationale and inferences concerning hypothesis tests are the same as before, only the computational methodology is different.

OBJECTIVES

- To test hypotheses about proportions arising from a single sample
- To test the hypothesis of the equality of proportions arising from two or more separate samples
- To place a confidence interval on a single population proportion
- To place a confidence interval on the difference in two population proportions

FOUNDATIONS

Data from a Single Sample

The statistical techniques presented in previous chapters were based on measurements of continuous variables. For a single population, the parameter was μ, which was estimated by the sample mean $\bar{Y}$. When the measurement on each subject in a sample is merely the presence or absence of a condition, the population parameter is P, the proportion with the condition in the population, which is estimated by p, the proportion with the condition in the sample.

A test of hypothesis about the population proportion P based on a sample proportion p is analogous to tests of hypothesis about μ as described in Chapter 4. In this instance, the null hypothesis is that P has a specified value P_0 rather than $\mu = \mu_0$.

The rationale for the χ^2 test is that if the true population proportion of a certain condition is P_0 and a sample of n subjects from the population is observed, then nP_0 of the total n subjects would be expected to display the condition. If the actual observed number of subjects with the condition in the sample is close to the expected value, then the null hypothesis that $P = P_0$ is not rejected. The procedure for computing the χ^2 statistic will be presented with examples in the methodology section.

The preceding discussion assumed that members of the population were classified according to the presence or absence of only one condition. Examples include the presence or absence of postoperative infection, a positive or negative diagnosis of carcinoma of the lung, and the success or failure of an organ transplant. This scheme also applies to a population with only two distinct conditions, say male and female, since sample subjects could be classified as "female" and "not-female."

A sample can also arise from a population whose members can be classified as having one and only one of three or more conditions. Examples include classification as Black, White, or Other; Stage I, Stage II, Stage III, or Stage IV of disease severity; and cause of death as one of the many (but finite) International Classification of Disease Codes. In this case, a true population proportion is hypothesized for each condition or attribute and compared to the observed sample proportions using the appropriate χ^2 statistic.

Data from Two or More Samples

The simplest case when comparing populations is that of two populations which are hypothesized to have the same population proportion of a single condition or attribute. Examples include the proportion of adult males with hypertension among black and white populations and the proportion of stillbirths among selected smoking mothers as compared to a matched sample of nonsmoking mothers. The null hypothesis is that $P_1 = P_2$ where P_1 is the true proportion in the first population and P_2 is the true proportion in the second population. The test statistic compares the difference in sample proportions, $\widehat{p_1} - \widehat{p_2}$, in a manner analogous to the comparison of $\bar{Y}_1 - \bar{Y}_2$ in Chapter 4.

When there are more than two populations involved, the null hypothesis is $P_1 = P_2 = \cdots = P_k$ and the test is analogous to the one-way analysis of variance. Examples are the proportion of students passing Part 1 of the National Board Exams for each of the medical schools in the southeast United States and the proportion of survivors from each of several groups of coronary artery disease patients.

The most general case is that of comparing several populations according to the proportions of members in three or more conditions within each population. An example would be the proportions of primarily circulatory, respiratory, or neurological diagnoses made among samples of patients of a general practitioner, an internist, and a family practitioner. The null hypothesis would be that the distribution (proportion) of patients among the diagnostic categories would be the same for the three physicians.

Each of the cases discussed in the two previous sections will be illustrated with numerical examples in the following section.

METHODOLOGY

Examples of Enumeration Data from a Single Sample

Data from a sample that has been drawn to estimate the proportion of individuals in the population displaying a certain attribute may be represented as follows:

	Observed
No. with attribute	n_1
No. without attribute	n_2
Total observations	$n = n_1 + n_2$

The estimate from the sample of the true population proportion, P, of individuals with the attribute is given by $\widehat{p} = n_1/n$.

When the true population proportion is P, then the expected number of observations displaying the attribute in the sample of size n is given by nP and the expected number without the attribute is given by $n(1 - P)$. In testing hypotheses about the value of the true population proportion, one assumes a value for P and then tests to see if this value is supported by the sample results. This is analogous to tests of hypothesis about μ described in Chapter 4.

The expected values together with the observed values may be displayed as follows:

	Observed	Expected
No. with attribute	n_1	nP
No. without attribute	n_2	$n(1 - P)$
Total	n	n

Note that $nP + n(1 - P) = n(P + 1 - P) = n$ so that the sum of the expected values is equal to the sum of observed values.

The χ^2 statistic for testing the null hypothesis that the true population proportion is P is computed as

$$\chi^2 = \sum \frac{(|\text{observed} - \text{expected}| - .5)^2}{\text{expected}}$$

where the straight lines bracketing the difference "observed-expected" means "disregard the sign and consider the difference to be positive" before subtracting the 0.5. The so-called correction factor of 0.5 is used when a single proportion is being tested.

The computed χ^2 is then compared to the tabulated value found in Appendix Table A for one degree of freedom and the stated significance level. If χ^2 computed is greater than or equal to χ^2 tabulated, then the null hypothesis is rejected. An example follows.

In a clinical trial to determine the preference for a new analgesic in relief of headache, 100 patients who suffered from chronic headaches were given the preparation. After a specified time interval each patient was asked whether he preferred the new preparation over the standard preparation he had been using. Of the 100 patients, 60 claimed to prefer the new analgesic. Based on this response, can it be said that the new preparation is preferred over the old?

The data in the above example may be displayed as follows:

	Observed
Preferred	60
Not preferred	40
	$n = 100$

The null hypothesis of interest is that the proportion of patients preferring the new preparation is the same as the proportion of patients who prefer the old. Stated formally,

$$H_0 : P = 0.5$$

If, indeed, the true population proportion of patients preferring the new drug is 0.5, then of the 100 patients in the sample, one would "expect" $nP =$

$(100)(.5) = 50$ patients to fall in this category. The number of patients actually observed in the sample was 60. As was the case in previous tests of hypothesis, this difference may have arisen because of sampling error or the true population proportion may not in fact be equal to 0.5.

Table 7.1 gives the observed values along with the expected values for both categories. The χ^2 statistic is computed as

$$\chi^2 = \sum \frac{(|\text{observed} - \text{expected}| - .5)^2}{\text{expected}}$$

$$= \frac{(|60 - 50| - .5)^2}{50} + \frac{(|40 - 50| - .5)^2}{50}$$

$$= \frac{9.5^2 + 9.5^2}{50} = 3.61$$

From Table A we find that the tabulated χ^2 for $\alpha = 0.05$ and one degree of freedom is 3.84. Since the calculated χ^2 is less than the tabulated χ^2, the null hypothesis of equal preference cannot be rejected. There is not enough evidence to support a claim that the new preparation is preferred for relieving chronic headaches.

The rationale for the χ^2 test is that if the true population proportion is P then, except for sampling variation, nP should be "close to" n_1 (i.e., p is "close" to P). When observed and expected values are close together, then the computed χ^2 value is small, and the null hypothesis is not rejected. When observed and expected values are far apart, the computed χ^2 value is large, leading to rejection of the null hypothesis.

In the preceding example the null hypothesis was $P = 0.5$. This is not always the case. For example, a particular operative technique may have associated with it a success rate of 80 percent. A surgeon wishes to test whether a new technique improves success rate. In this case the data from a

Table 7.1. Observed and expected number of patients experiencing relief or no relief of symptom

	Observed	Expected
Preferred	60	50
Not preferred	40	50
	100	100

sample of patients receiving the new procedure would be presented as

Observed

Successful	n_1
Had complications	$\dfrac{n_2}{n}$

The null hypothesis would be that the true proportion of successes, P, is equal to 0.8.

The previous example may be extended to the case in which a single sample is classified according to multiple attributes. The data may be displayed in tabular form as follows:

Observed

	1	n_1
	2	n_2
Attribute	⋮	⋮
	c	n_c
Total		$n = \sum n_i$

The null hypothesis specifies that the true population proportions for the categories are $P_1, P_2, \ldots, P_{c-1}$. Only $c - 1$ proportions have to be specified since P_c may be obtained by subtraction, that is, $P_c = 1 - (P_1 + P_2 + \cdots + P_{c-1})$.

Since there are $c - 1$ independent expected values (nP_c may be obtained by subtraction), the degrees of freedom associated with χ^2 is $c - 1$. For three or more proportions, df is greater than one so the correction factor is not needed.

The "closeness" of the observed and expected values is measured by again calculating the χ^2 statistic,

$$\chi^2 = \sum \frac{(\text{observed} - \text{expected})^2}{\text{expected}}$$

The null hypothesis is rejected when the value for the calculated χ^2 exceeds the value for χ^2 ($df = c - 1$) obtained from Table A in the Appendix for a specified significance level.

A single sample of female patients classified according to type of contraceptive used is an example of hypothesis about more than one proportion. In a large medical school-affiliated hospital, all women from age 15–45 who were discharged with a diagnosis of idiopathic thromboembolism were identified and then classified as to type of contraceptive used. The data are given in Table 7.2. Is there a difference in contraceptive usage among women suffering from thromboembolic disease? The null hypothesis in this example would be

$$P_1 = P_2 = P_3 = P_4 = 0.25$$

Table 7.2. Contraceptive usage among thrombo-embolic women

	Observed
Oral contraceptive	30
IUD	25
Diaphragm	20
None	25
	100

As before, expected values for each attribute must be obtained. If the true proportion of thromboembolic women using oral contraceptives is $P_1 = 0.25$, then we would expect $nP_1 = (100)(.25) = 25$ women in the sample to fall in this category. This is compared to the 30 women actually observed in the category. The remaining expected values are shown below.

	Observed	*Expected*
Oral contraceptive	30	25
IUD	25	25
Diaphragm	20	25
None	25	25

The calculated χ^2 is

$$\chi^2 = \sum \frac{(\text{observed} - \text{expected})^2}{\text{expected}}$$

$$= \frac{(30 - 25)^2}{25} + \frac{(25 - 25)^2}{25} + \frac{(20 - 25)^2}{25} + \frac{(25 - 25)^2}{25}$$

$$= 1 + 0 + 1 + 0 = 2$$

For $\alpha = 0.05$ and $df = 3$, the value from Table A is 7.81. The null hypothesis of equal population proportions, therefore, cannot be rejected. Contraceptive usage cannot be shown to be different among thromboembolic women.

Enumeration Data from Separate Samples

Two samples. The following tabulation represents 60 patients assigned to a new treatment as compared to 40 patients receiving the standard treatment for a particular disease.

	Improved	Not Improved	
New treatment	40	20	60
Standard treatment	25	15	40
	65	35	100

The various totals in the table have specific names. The *total* number of observations is 100; the *marginal total* of patients under the new treatment is 60; the marginal total of patients showing improvement is 65; the *cell total* of patients not improved under the standard treatment is 15. The other totals are named in a similar manner.

The research question is whether the new treatment shows a significantly higher percentage of improved patients or whether the proportion of improved patients is statistically the same for both treatments. In terms of the null hypothesis, this may be written

$$H_0 : P_1 = P_2 = P$$

where P_1 is the true proportion of patients having improved status after receiving the new treatment and P_2 is likewise the true proportion of patients having improved status after receiving the standard treatment. Estimates of the population parameters are obtained from cell and marginal totals:

$$\hat{p}_1 = \frac{40}{60} = 0.67$$

and

$$\hat{p}_2 = \frac{25}{40} = 0.63$$

Are p_1 and p_2 close enough in value for one to assume that $P_1 = P_2 = P$? To answer this we must again calculate the "expected" count for each category. Recall that the null hypothesis claims equal improvement rates for the new and standard treatments. Then the best estimate of the unknown population improvement rate P would be the proportion improved independent of which treatment was used:

$$p = \frac{\text{Total number improved}}{\text{Total patients}} = \frac{65}{100} = 0.65$$

Among the 60 people receiving the new treatment, we would expect $60(\frac{65}{100})$ or 39 people to improve if the null hypothesis is true. Correspondingly, we would expect $40(\frac{65}{100})$ or 26 of those receiving the standard treatment to improve.

Applying the same reasoning, we estimate the common "not improved" rate under the null hypothesis by

$$1 - p = \frac{\text{Total number not improved}}{\text{Total patients}} = \frac{35}{100}$$

Table 7.3.

	Improved	*Not Improved*	
New treatment	40(39)	20(21)	60
Standard treatment	25(26)	15(14)	40
	65	35	100

Based on this estimate we would, therefore, expect $60(\frac{35}{100})$ or 21 of the new treatment group and $40(\frac{35}{100})$ or 14 of the standard treatment group to fall in the "not improved" category if the null hypothesis is true. Note that the expected value for each cell is given by the product of the marginal totals corresponding to the cell divided by the total number of observations. Table 7.3 shows both the observed and expected values. The expected values are given in parentheses. Verify that row totals and column totals are the same for both observed and expected values. Since the marginal totals are known, one needs only to calculate one of the expected values and obtain the others by subtraction. Since for this table there is only one independent expected value (i.e., all others obtained by subtraction), the χ^2 has associated with it only one *df*.

The χ^2 value is determined by the formula

$$\chi^2 = \sum \frac{(|\text{observed} - \text{expected}| - .5)^2}{\text{expected}}$$

$$= \frac{(|40 - 39| - .5)^2}{39} + \frac{(|20 - 21| - .5)^2}{21}$$

$$+ \frac{(|25 - 26| - .5)^2}{26} + \frac{(|15 - 14| - .5)^2}{14}$$

$$= \frac{(.5)^2}{39} + \frac{(.5)^2}{21} + \frac{(.5)^2}{26} + \frac{(.5)^2}{14} = 0.046$$

where the correction factor is used since χ^2 has only one *df*. The tabulated χ^2 from Table A for $\alpha = 0.05$ and $df = 1$ is 3.84. Since the calculated χ^2 is less than the tabulated value, we cannot say that rate of improvement is different for the two groups.

A comparison among more than two populations on a single condition is a straightforward extension. Over a 2-year period a total of 300 people with hypertension were treated at a clinic. Each patient was given one of three drugs (drug A, drug B, or drug C). At the end of 3 months of therapy each patient's condition was categorized as "no change" or "improved." The number falling into each category is given in Table 7.4.

Table 7.4. Results of a 2-year
hypertension study

	No Change	Improved	
Drug A	20	20	40
Drug B	40	60	100
Drug C	40	120	160
	100	200	300

The physician is interested in determining if there is any difference between the drugs. If the efficacy of the three drugs is the same, then the proportions of people showing improvement will be the same except for random variation. That is, the proportions improved are the same for each drug:

$$H_0 : P_A = P_B = P_C = P$$

As before, the expected values are found by using the marginal totals. The expected value of any cell is found by multiplying its row total by its column total and dividing by the total number of observations.

First row, first column: $\dfrac{(40)(100)}{300} = 13.33$

First row, second column: $\dfrac{(40)(200)}{300} = 26.67$

$\vdots \qquad\qquad\qquad \vdots$

Last row, last column: $\dfrac{(160)(200)}{300} = 106.67$

Table 7.5 gives observed and expected frequencies for this example. The calculated χ^2 is

$$\chi^2 = \sum \frac{(\text{observed} - \text{expected})^2}{\text{expected}}$$

$$= \frac{(20 - 13.33)^2}{13.33} + \frac{(20 - 26.67)^2}{26.67}$$

$$+ \frac{(40 - 33.33)^2}{33.33} + \frac{(60 - 66.67)^2}{66.67}$$

$$+ \frac{(40 - 53.33)^2}{53.33} + \frac{(120 - 106.67)^2}{106.67} = 12.005$$

Table 7.5. Expected values for the data in Table 7.4

	No Change	Improved	
Drug A	20(13.33)	20(26.67)	40
Drug B	40(33.33)	60(66.67)	100
Drug C	40(53.33)	120(106.67)	160
	100	200	300

Since there are two independent expected values (all others may be obtained by subtraction), there are two *df* associated with the χ^2 value. The degrees of freedom may also be determined by the formula

$$df = (\text{rows} - 1)(\text{columns} - 1)$$

$$= (3 - 1)(2 - 1) = 2$$

The tabulated χ^2 for $\alpha = 0.01$ and $df = 2$ is 9.21. The null hypothesis of equal proportion in the condition classification is rejected. We may conclude that there is a difference in efficacy among the three drugs.

By an obvious extension the case of multiple conditions for more than two samples can be analyzed. In order to study anxiety levels among medical students, 150 freshmen, 135 sophomores, 125 juniors, and 100 seniors were selected and classified according to anxiety level. The data is presented in Table 7.6. The null hypothesis may be written

$P_{11} = P_{21} = P_{31} = P_{41}$ (proportion with low anxiety equal among classes)

$P_{12} = P_{22} = P_{32} = P_{42}$ (proportion with moderate anxiety equal among classes)

$P_{13} = P_{23} = P_{33} = P_{43}$ (proportion with high anxiety equal among classes)

Table 7.6. Anxiety levels among medical students

	Low(1)	Moderate(2)	High(3)	
Freshmen(1)	50	57	43	150
Sophomores(2)	58	57	20	135
Juniors(3)	45	56	24	125
Seniors(4)	22	45	33	100
	175	215	120	510

In words, we wish to determine if the proportions of students falling in the different anxiety classifications are constant from class to class. Expected values are determined as before.

First row, first column: $\dfrac{(150)(175)}{510} = 51.47$

First row, second column: $\dfrac{(150)(215)}{510} = 63.24$

$\vdots$ $\qquad\qquad$ $\vdots$

Last row, last column: $\dfrac{(100)(120)}{510} = 23.53$

Table 7.7 gives observed and expected frequencies.

Table 7.7. Expected frequencies for the data in Table 7.6

	Low	Moderate	High	
Freshmen	50(51.47)	57(63.24)	43(35.29)	150
Sophomores	58(46.32)	57(56.91)	20(31.76)	135
Juniors	45(42.89)	56(52.70)	24(29.41)	125
Seniors	22(34.31)	45(42.16)	33(23.53)	100
	175	215	120	510

$$\chi^2 = \sum \frac{(\text{observed-expected})^2}{\text{expected}}$$

$$= \frac{(50 - 51.47)^2}{51.47} + \frac{(57 - 63.24)^2}{63.24} + \frac{(43 - 35.29)^2}{35.29}$$

$$+ \frac{(58 - 46.32)^2}{46.32} + \cdots + \frac{(33 - 23.53)^2}{23.52} = 19.4$$

The degrees of freedom associated with χ^2 is

$$df = (r - 1)(c - 1) = (4 - 1)(3 - 1) = 6$$

From Table A the tabulated χ^2 for $\alpha = 0.01$ and $df = 6$ is 16.8. Since the calculated χ^2 is greater than the tabulated χ^2, we reject the null hypothesis of equal proportions and conclude that stress levels are different across classes.

Confidence Intervals on Population Proportions

Often, it may be of interest to estimate a population proportion, P, by means of a confidence interval when the experimental situation allows classification according to only two possible outcomes. Examples include the proportion of patients who show improvement when the patients have been classified as "improved" or "not improved," the proportion of "smokers" or "nonsmokers," or the proportion of individuals who have or do not have a specified disease. Fortunately, in order to determine a confidence interval for p, we can take advantage of a mathematical relationship between the z statistic and the χ^2 statistic with one degree of freedom. The relationship is that $z^2 = \chi^2$ ($df = 1$). For example, the tabulated z value for $\alpha = .05$ (two-tailed) is ± 1.96, and the one degree of freedom χ^2 value for $\alpha = .05$ is $(\pm 1.96)^2 = 3.841$. The reader should verify this relationship for $\alpha = .01$.

It may be recalled from previous chapters that the general form of a confidence interval is

Estimate $\pm$ (table value) $\times$ (standard error of estimate)

An estimate, $\hat{p}$, of the population proportion, P, is obtained by the formula

$$\hat{p} = \frac{\text{number in the sample with (or without) the attribute}}{\text{total number in the sample}}$$

Without proof, we define the standard error of the estimate to be

$$\sqrt{\frac{\hat{p}(1 - \hat{p})}{n}}$$

Making use of the relationship between z and χ^2, we may write the confidence interval in terms of z as

$$\hat{p} \pm z_{(1-\alpha)/2} \sqrt{\frac{\hat{p}(1 - \hat{p})}{n}}$$

Example. Consider the data shown in Table 7.1. A 95 percent confidence interval on the true proportion of patients who prefer the new preparation is

$$\frac{60}{100} \pm 1.96 \sqrt{\frac{\left(\frac{60}{100}\right)\left(\frac{40}{100}\right)}{100}}$$

$$[.50, .70]$$

We may be 95 percent confident that the interval $[.50, .70]$ covers the true proportion of individuals who prefer the new procedure.

Often, when placing a confidence interval on a single population proportion, it may turn out that the lower limit is a negative number. In such cases, it is common practice to record the lower limit as 0.

In addition to placing a confidence interval on a single population proportion, p, we may wish to determine a confidence interval for the true difference in population proportions, $P_1 - P_2$. A $100(1 - \alpha)$ percent confidence interval for $P_1 - P_2$ is given by

$$(\hat{p}_1 - \hat{p}_2) \pm (\text{table value}) \sqrt{\frac{\hat{p}_1(1 - \hat{p}_1)}{n_1} + \frac{\hat{p}_2(1 - \hat{p}_2)}{n_2}}$$

Example. Consider the data shown in Table 7.3. A 99 percent confidence interval on the difference in proportion of patients showing improvement for the new and standard treatments is

$$\left(\frac{40}{60} - \frac{25}{40}\right) \pm 2.58 \sqrt{\frac{\left(\frac{40}{60}\right)\left(\frac{20}{60}\right)}{60} + \frac{\left(\frac{25}{40}\right)\left(\frac{15}{40}\right)}{40}}$$

$$.042 \pm 2.58 \sqrt{\frac{(.67)(.33)}{60} + \frac{(.625)(.375)}{40}}$$

$$[-.21, .29]$$

where the proportion showing improvement for the new treatment is $\hat{p}_1 = 40/60$ and the proportion showing improvement for the standard treatment is $\hat{p}_2 = 25/40$. We are 99 percent confident that the interval $[-.21, .29]$ covers the true difference in proportion showing improvement for the two populations.

Testing Hypotheses About Specific Differences Following the Overall χ^2 Analysis

As was the case with a significant F test in the analysis of variance involving more than two groups, a significant overall χ^2 test with more than one degree of freedom indicates that there is a difference in proportions but does not indicate which proportions are different. For the special case of $r \times 2$ contingency tables, there is a straightforward application of the Bonferroni procedure as described in Chapter 6. Suppose for the data in Table 7.4, which is a 3×2 contingency table, drug A represents a standard treatment. The investigator is interested in comparing drug A to both drug

B and drug C with respect to proportion of patients showing improvement. The 2×2 contingency table for comparing drug A to drug B is shown in Table 7.8. The corrected χ^2 for this comparison is 0.79 with one degree of freedom. The reader may verify that the corrected χ^2 value for testing drug A against drug C is 8.37 with one degree of freedom.

Table 7.8. Results of 2-year hypertension study for drug A and drug B

	No Change	Improved	
Drug A	20(17.14)	20(22.86)	40
Drug B	40(42.86)	60(57.14)	100
	60	80	140

In order to preserve a per experiment error rate of $\alpha = .01$, we compare the two calculated χ^2 values with a tabulated $\chi^2_{.01/2}$ or $\chi^2_{.005}$ value.

Since we have only single degree of freedom comparisons, we may use the relationship $z^2 = \chi^2$ (1 degree of freedom) discussed in the preceeding section. Here, we look up the z value for $(1 - .005)/2 = .4975$. We have,

$$\chi^2_{.005} = z^2_{.4975} = (2.81)^2 = 7.90$$

Thus, we may conclude that drug A is not different from drug B, but drug A is different from drug C with an overall per experiment error rate of $\alpha = .01$.

Small Expected Frequencies

When expected values are small, the validity of the χ^2 tests are questionable. Much controversy exists as to what constitutes "small." Some statisticians feel that the expected count in each category should be at least five, whereas others feel that the test is valid so long as expected frequencies are greater than one. A possible solution is to combine adjacent categories when one or more expected frequencies are less than one.

Summary of χ^2

	Single Sample	Two Samples	More Than Two Samples
Single attributes	$\chi^2 = \sum \dfrac{(\lvert 0-E\rvert - .5)^2}{E}$	$\chi^2 = \sum \dfrac{(\lvert 0-E\rvert - .5)^2}{E}$	$\chi^2 = \sum \dfrac{(0-E)^2}{E}$
	$df = 1$	$df = 1$	$df = (r-1)(c-1)$

	Single Sample	*Two Samples*	*More Than Two Samples*
Multiple attributes	$\chi^2 = \sum \dfrac{(0-E)^2}{E}$	$\chi^2 = \sum \dfrac{(0-E)^2}{E}$	$\chi^2 = \sum \dfrac{(0-E)^2}{E}$
	$df = (c-1)$	$df = (c-1)$	$df = (r-1)(c-1)$

Note that when $df = 1$, a "correction factor" of 0.5 is subtracted from $|0 - E|$.

PROBLEMS

1. If the reported national success rate for a particular operation is 90 percent, what number of unsuccessful operations would you expect to find in a review of 100 records in your hospital if your rate is the same as the national average? Suppose you found records for 15 unsuccessful operations among the 100 surveyed. Does this experience reflect the national experience? (Use $\alpha = 0.05$.) Place a 95 percent confidence interval on the true population proportion of successful operations.

2. In a study on the effect of dental hygiene instruction on dental caries in children, 100 children were randomly selected; 50 received instruction and 50 did not. At the end of a 6-month period, the number of new cavities for each child was recorded. The data is given below.

	Number of New Cavities			
	0-1	*2-3*	*4-5*	*Total*
Instruction	30	15	5	50
No instruction	20	15	15	50
				100

Using the 5 percent level of significance, can it be said that there is an association between instruction in dental hygiene and number of new cavities at the end of a 6-month period?

3. Forty females using oral contraceptives and 60 females using other contraceptive devices were randomly selected, and the number of hypertensive for both groups were recorded as follows:

	Total	*No. Hypertensive*
Oral contraceptives	40	8
Other	60	15

Test the hypothesis that the proportion of patients with hypertension is the same for the two groups. (Use $\alpha = 0.01$.) Construct a 99 percent confidence interval on the true difference in proportion hypertensive for the two groups.

4. In order to examine the effects of tranquilizers and stimulants on driving skills, 150 people selected at random were given either a stimulant, a tranquilizer, or an identically appearing placebo. After receiving the medication, the participants were given a battery of coordination and reaction time tests and the number of "mistakes" for each was recorded. The following table gives total number of mistakes for the entire battery of tests for the three groups:

| | Total Number of Mistakes | | | |
	0–5	6–10	11–15	Total
Stimulant	10	20	20	50
Depressant	5	15	30	50
Placebo	25	15	10	50
				150

Based on this data, can it be concluded that the proportion of mistakes is different for the three groups? (Use $\alpha = 0.05$.)

5. Freshmen medical students were asked to rate a particular core course according to level of difficulty of the material presented in lectures. The following table shows this response cross-tabulated with grade received in the course:

| | Perceived Difficulty of Course | | | |
| Class Rank | Not | Intermediate | Very | |
(Final Grade)	Difficult	Difficulty	Difficult	Total
Upper 1/3	10	40	17	67
Middle 1/3	20	35	12	67
Lower 1/3	21	15	30	66
				200

Is there an association between perceived difficulty of the course and class rank? (Use $\alpha = 0.05$.)

6. Thirty patients receiving a tricyclic antidepressant for management of their depressive states were given a psychomotor skill test. Their overall response was rated either satisfactory or not satisfactory. Another group of 30 patients receiving the same medication were given alcohol in addition to the antidepressant and their psychomotor skills were evaluated. Twelve of the patients receiving only the antidepressant performed satisfactorily, whereas 8 of the patients receiving both the antidepressant and alcohol performed satisfactorily. Is there a difference in psychomotor performance for the two groups? (Use $\alpha = 0.05$.)

7. One hundred institutionalized schizophrenic patients were classified according to the season of the year in which they were born.

Season of Birth Among 100
  Schizophrenic Patients

	Observed
Fall	20
Winter	35
Spring	20
Summer	25
	100

Based on the above data, can it be said that there is a difference in season of birth among institutionalized schizophrenic patients? (Use $\alpha = 0.05$.)

8. Fifty impatient insomniacs were treated with equipotent doses of three currently used hypnotics and a placebo. Of the 50 patients in each of the treatment groups, those who felt that they still did not get a "good night's sleep" were classified according to their perceived reasons for not sleeping well.

Reason Given for Not Sleeping Well	Hypnotic A	Hypnotic B	Hypnotic C	Placebo
Restlessness	2	5	4	6
Woke too early	5	7	5	10
Trouble getting to sleep	8	4	6	14
Bad dreams	3	4	3	2
Other	2	5	2	3
Total	20	25	20	35

Is there an association between reasons given for not sleeping well and type of medication received by the patient? (Use $\alpha = 0.01$.)

CHAPTER 8

TYPES OF CLINICAL AND EPIDEMIOLOGICAL STUDIES

OVERVIEW

The statistical techniques introduced in the preceding chapters have been placed in the context of specific questions about the numerical interpretation of medical or health data without a detailed discussion of the circumstances surrounding their collection. As might be anticipated, the experimental design (if any) that is followed in generating and collecting the data is extremely important for relating the statistical tests to the population(s) being studied. Medical studies divide themselves naturally into comparisons of drugs, procedures, or therapeutic regimens, generally under the heading of *controlled clinical trials*, and studies of the association of (potentially) causal factors of diseases in populations under the heading of *observational epidemiological studies*.

The methods and procedures of controlled clinical trials and observational epidemiological studies are described below.

OBJECTIVES

- To distinguish between observational and experimental studies
- To list the characteristics of a controlled clinical trial
- From a description of the study, to decide whether it is prospective, retrospective, or cross-sectional in nature
- To list the advantages and disadvantages of prospective, retrospective, and cross-sectional studies.

FOUNDATIONS

Introduction

Medical studies can usually be categorized as observational or experimental. An *observational study* is one in which the investigators do not directly control which population units are observed nor do they control the condition or factors to which the units are exposed. Thus, observational studies utilize ongoing, natural phenomena to attempt to relate disease to possible associated factors. In an *experimental study*, the investigators control which population units enter the study, which factors at what levels the population units are exposed to, and the conditions under which the experiment is carried out.

From the definition of experimental studies it is clear that they would be almost impossible to carry out on large, diverse population groups. This method generally finds its best application in clinical studies of disease outcome under various treatment regimens. Observational studies, on the other hand, find their application in the investigation of the determinants, distribution, and prevention of disease among human populations, which is the province of epidemiology. Sometimes situations arise spontaneously which have all the earmarks of an experimental study. Such situations are called *natural experiments* and provide valuable epidemiological information.

Clinical Trials

Controlled clinical trials. The prototype of clinical experimental studies is the controlled clinical trial. Usually designed to select the best of two or more treatments, these studies are controlled in the sense that only patients who meet rigorous criteria of eligibility and comparability are enrolled. Care is taken that patients in each treatment group are as similar as possible to patients in every other treatment group at the outset with respect to any variables that might interfere with the interpretation of the possible differences among treatments. Further, the experiment is carried out in such a fashion that the conduct of the study is similar for each treatment group.

One procedure is to randomly assign eligible patients to the treatment groups to avoid any bias on the part of the investigators concerning which patient should receive what treatment. Then a further safeguard is taken by not allowing either the evaluator of the treatment outcome or the patient to know which treatment the patient is receiving. This process yields a *double-blind* study. This is especially important when one "treatment" is an inactive preparation, or *placebo*.

The overall goal of a controlled clinical trial is to define treatment groups and to conduct the study in such a way that the only difference among groups is attributable to differences in the effects of the treatments.

Uncontrolled clinical trials and case reports. Studies that reflect clinical observations on a series of routinely available patients treated in a certain way without a concurrent comparison series are frequently reported. Such studies are useful but lack the definition of controlled clinical trials. Likewise, reports on cases that have been selected because of some commonality among them are useful to indicate possible directions of research but are limited by lack of experimental control and defined comparison groups.

Community trials. Sometimes complete communities are the subject of experimental studies in which the "treatments" are applied to the various communities and the outcomes are compared on the basis of summary statistics for the whole community. Clearly, there can be no randomization of individuals, and the conduct of such a study is not as easily controlled as are clinical trials. Nevertheless, community trials often yield valuable information.

Observational Epidemiological Studies

As mentioned earlier, the primary function of epidemiology is to study the determinants and distribution of disease in human populations with a view toward subsequent prevention. The three major types of observational studies in epidemiology are the prospective, the retrospective, and the cross-sectional.

Prospective studies. Prospective studies, also called cohort or longitudinal studies, begin with two or more groups (or cohorts) of persons all free of the disease under study but with each group having had a differential exposure to a possible etiologic factor. The cohorts are then followed over time (longitudinally) to measure whether the disease in question appears at a different rate among the various groups during the course of the study.

If the rate of newly diagnosed disease is higher among those exposed to the factor, or to higher levels of the factor, this may be evidence that the factor is associated with the development of the disease.

In the prospective study, as in all observational studies, the lack of experimental control makes it incumbent on the investigators to ensure the comparability of the groups being studied. Every effort must be made to show that the only important difference among groups with respect to outcome is the exposure to the possibly causal factor. This includes consideration of demographic factors (e.g., age, sex, race, socioeconomic status), environmental factors (e.g., pollution, noise, climate), and life-style factors (e.g., smoking, use of alcohol and drugs, nutrition), among others. Since many differences among groups can be adjusted statistically, the investigator exerts *statistical control* over possible confounding variables rather than *experimental control.* The more comparable the groups at the outset, the more dependable are the results of the study.

The advantages of prospective studies are that (1) they allow for rigorous criteria for inclusion in the study in terms of documenting the presence or absence of the factor under study and the documented absence of the disease; (2) they permit identification of new cases of the disease by rigid criteria; and (3) they allow for direct measurement of the number of cases of the disease that develop in the various study groups during the study.

The major disadvantages of prospective studies are that (1) they require follow-up on each person over the complete time course of the study; (2) they are time consuming and expensive; and (3) large numbers of subjects are needed if a rare disease is being studied.

A major problem in prospective studies is persons lost to follow-up. Statistical methods to address this problem will be introduced in the next chapter.

Retrospective studies.   Retrospective studies, also called case-control studies, begin with finding a group of subjects with the disease under study (cases) and a comparable group without the disease (controls) and then looking back (retrospectively) to determine whether there has been differential exposure to a possible etiologic factor at some point in the past. Thus, retrospective studies begin with persons with the disease and look back in time for the presence of the factor, while prospective studies begin with the factor and look forward in time to the development of the disease. If the possible risk factor occurs more frequently or at a higher level among the cases than among the controls, this may be evidence that the suspected factor is associated with the disease.

Retrospective studies, by definition, require the prior diagnosis of the disease before a case is identified. Consequently, hospital medical records, clinical records, insurance records, death certificates, and the like are often the source of case identification. Sometimes, the record is also the source of information on exposure to the suspected factor (e.g., employment history, residence). The criteria for the definition of a case, and the exclusion of a potential control from being a case, is a major issue in case-control studies since such decisions have already been made, usually by persons not aware that the records will be used for research. This is important because the shift of a few subjects from case to control, or vice-versa, could completely change the results.

The correctness of the data on exposure to the factor under study is also of concern. Not only must records be examined carefully, but even information obtained from living cases or families of cases is subject to recall errors.

The major problem, in the view of many investigators, is the proper definition of a comparison group for a retrospective study. This is so because it is extremely difficult to ascertain whether two groups were similar in the past and that they have had comparable life experiences except for exposure to the factor under study. Often matched pairs of cases and controls, where the matching is done on variables that might interfere with the interpretation of the data, are used in an attempt to make the study groups more comparable. Even so, there are still insidious problems such as the rate at which cases with the factor and controls with the factor are included in the study, not reflecting the proportion of cases and controls with the factor in the population. This is known as *Berkson's fallacy* and is to be watched for especially in the use of hospital records where there might be differential hospitalization rates.

In spite of these problems, which revolve around the correct identification of disease and exposure, the retrospective study is an extremely useful and popular tool for epidemiological research.

The major advantages of a retrospective study are that it can be done quickly and inexpensively, and that large numbers of cases can be studied even for rare diseases. A properly designed and executed retrospective study can be a major step in the elucidation of associations between suspected factors and disease.

In summary, the advantages of retrospective studies are that (1) they may be readily carried out using existing data; (2) they are usually not expensive, especially in comparison to other alternatives; and (3) they can be carried out on rare diseases because they can use cases accumulated over a long period of time. The major disadvantages of retrospective studies are (1) the problems associated with choosing a proper control group; (2) the accuracy with which cases are known to have the disease and controls are known to be free of the disease; and (3) the accuracy with which cases are known to have been exposed to the factor and controls are known not to have been exposed to the factor.

Cross-sectional studies.   A cross-sectional study is similar to a retrospective study in that one group has the disease of interest while the comparison group is free of the disease. However, the cross-sectional approach differs from both the retrospective and prospective approaches in that it looks at the simultaneous occurrence of the study disease and study factor in the present, hence the name cross-sectional for one point in time.

The concurrent nature of the disease and the "factor" and their measurement during a single time period (usually short) makes it impossible to determine which came first. This is especially true for variables that could possibly have been influenced by the (latent) presence of the disease. In addition to this shortcoming, the cross-sectional study has problems and advantages similar to retrospective studies.

Association and Causation

Generally, the aim of both clinical and epidemiological studies is to relate "factors" to their "effects" in a causal fashion. For individual patients one would like to say at the end of a clinical trial whether an improvement in condition will follow (be caused by) the treatment. For populations, which are the targets of epidemiological studies, one would like to say at the end of the investigation that the improvement of a certain health index will follow (be caused by) the removal of the risk factor. In this sense one aims to show a directly causal statistical association between the factor and the effect. It is rare for factors and effects to be (mathematically) causally related in that every subject shows the effect following exposure to the factor; thus the phrase "causal statistical association."

Since clinical trials are under experimental control and exposure to the factor can be directly manipulated, direct causal statistical associations can be quite clear cut and readily interpretable. Such is not the case for observational epidemiological studies; therefore, great care must go into defining the nature of the relationship, if any, between factor and effect at the conclusion of such studies.

In general, statistical associations may be classified as *spurious*, *indirect*, and *causal*. Spurious (incorrect) associations arise from some defect in the study, either in its execution or design. The most common cause of spurious associations are biases in the selection of cases or controls and biases in the method of collecting data. Indirect associations arise when one or more variables not covered in the study link the factor and the disease. Note that it is not necessary that the unknown factor(s) be a causal agent.

The existence of causal associations should be entertained only if there is a strong statistical association, if exposure to the suspected factor precedes the effect, if the association is biologically plausible according to current knowledge, and if no other data or other interpretation offer viable alternatives. Also, where possible, a dose–response relationship (i.e., a higher exposure leads to a larger effect) is suggestive of a causal effect.

METHODOLOGY

Examples will be given of the various types of studies discussed above. Suppose we are concerned with the effects of sodium consumption (the "exposure" of interest) and problem pregnancies (the "disease" of interest) characterized by fetal distress and maternal illnesses secondary to elevated maternal blood pressure, the following represent a series of studies that might be performed to link this factor to this health event.

Case-series. Obstetricians reviewing their own experience of difficult pregnancies attributable to maternal high blood pressure might choose five cases particularly "characteristic" of the problem. They would describe the mothers in great detail (e.g., age, race, weight, parity, clinical presentation) and similarly the pregnancy. With case-series there is no measure of risk calculated; the conclusions might be simply the observation that all of the mothers had elevated sodium levels and reported high sodium diets.

Cross-sectional study. Observers of the case series, who wish to have a quick assessment of the association between the high sodium diet and pregnancy problems, would select a representative group of women having difficult pregnancies and a similarly representative group not having problem pregnancies. Each group would then be tested for sodium levels of the blood. The observers, although unable to indicate whether the high sodium levels preceded or resulted from the problem pregnancy, would be able to demonstrate the association of the factor and the "disease."

Retrospective (observational) study. If investigators would then choose to further study this association (e.g., in order to determine a measure of risk for problem pregnancies due to the high sodium exposure), they would propose an observational, retrospective study. Selecting a "case" group of persons having problem pregnancies and a "control" group of persons not having these difficulties, they would obtain dietary and other relevant histories to determine the subjects' exposure to sodium before and during their pregnancies. Using dietary assessment (quantification) methods, they may be able to demonstrate that cases use much more salt on their foods than do controls.

Natural experiment. In studies of causal association, it is always good to demonstrate a gradient effect. Let us suppose an investigator sought and identified three ethnic communities, one using characteristically high salt diets, another a moderate salt diet, and the third (for religious reasons) using a low salt diet. The investigator could then survey these populations for the rates of problem pregnancies in each. The data from such a study may involve interviewing mothers, reviewing hospital charts, or identifying currently pregnant subjects. The data from the study would provide a gradient effect for evaluation of the consistency of the exposure and its risk of problem pregnancies. This is a natural experiment as there is no assignment of exposure and yet the essential exposures are represented.

Prospective study. Assume that the preceding natural experiment demonstrated a gradient effect. Next an experiment would be considered to follow subjects at specified levels of risk. Three obstetric practices might be selected: one that would not restrict diet at all; one that would restrict diet to moderate salt consumption; and a third that would adhere to a strict salt-free diet. All new pregnancies seen in these practices for a year would be followed to term. Assignment of persons to the diets is independent of the investigator (e.g., the people choose their own physicians). Measures of the risk of high, moderate, and low sodium consumption would now be based on the incidence of problem pregnancies in each group.

Clinical trial. We might wish to evaluate a salt substitute (potassium chloride) to reduce the problems involved in difficult pregnancies. In several obstetric practices, problem pregnancies could be identified (for example) in their first trimester. Subjects would be randomly assigned to no change in diet, salt-free diets, or to a diet using the salt substitute. They would be followed to term to assess the effect of the different treatment regimens in reducing problems in the further course of pregnancy. As investigators controlled assignment to each of the treatment regimens, they could stratify the groups to be "balanced" for suspected confounding factors (e.g., age, race, parity, weight). For further control of bias, the physicians treating the pregnant women could be kept unaware of their dietary regimens so that their evaluation of pregnancy problems would be blinded. In a single-blind study, the pregnant women would not know which diet they were receiving. In a double-blind study, both the women and their physicians would not know their dietary assignment. Assuming that the results of the study showed a reduction in the rate of complications, a larger trial would be undertaken to study whether problem pregnancies may be prevented by regulating sodium intake of pregnant women.

Community trial. In a community whose rate of problem pregnancies was known to be high, newly pregnant women would be assigned to regimens using the new salt substitute. This study is prospective, since the persons are being followed over time; yet it differs from the clinical trial in that the regimens are not intended as therapy, but rather as prophylaxis. The emphasis here is on group experience, as compared to the individual experience of the clinical trial. The reduction in the incidence of the disease (in this case, problem pregnancies) *in the population* would provide a measure of the efficacy of this factor, and thus an indication of its relation to the disease.

CHAPTER 9

ASSESSING RISK FROM OBSERVATIONAL EPIDEMIOLOGICAL STUDIES

OVERVIEW

In the preceding chapter, the various designs appropriate for observational epidemiological studies were presented without specifying the statistical analyses that follow the data collection. As we will see below, one measure (chi-square) will be used to determine the *significance* of the association between the disease and the suspected etiologic factor, while another measure (*odds ratio*) will be used to assess the *strength* of such an association. This is similar to the interpretation of regression and correlation, in which a significant regression indicates that a relationship exists and the correlation coefficient is used to indicate the strength of the relationship.

OBJECTIVES

- To perform and interpret chi-square analysis of four-fold contingency tables arising from cross-sectional, prospective, and retrospective studies
- To compute and interpret the odds ratio for cross-sectional, prospective, and retrospective studies

FOUNDATIONS

The theoretical background for the chi-square analysis of contingency tables has been presented in Chapter 7. The odds ratio and its interpretation will be discussed below.

Table 9.1. Prevalence of hypertension among women currently using oral contraceptives

Using Contraceptives	*Hypertensive* *Yes ($= B$)*	*No ($= \bar{B}$)*	*Total*
Yes ($= A$)	a	b	$a + b$
No ($= \bar{A}$)	c	d	$c + d$
Total	$a + c$	$b + d$	$\underline{a + b + c + d}$

Suppose that a team of investigators carries out a cross-sectional study to assess the association, if any, between the current use of oral contraceptives (factor A) and the current presence of hypertension (factor B). The data would be displayed as in Table 9.1.

In Table 9.1 the total number of subjects, $a + b + c + d$, is underlined to indicate that the data are from a cross-sectional study of a random sample of $a + b + c + d$ women who have been categorized according to the four cells of the (four-fold) contingency table.

These data can be used to calculate certain probabilities of interest (see Chapter 2). In particular,

$P(B|A)$ = probability that the disease (hypertension) is present given that the possible etiologic factor (oral contraceptive use) is present

$$= a/(a + b).$$

Further,

$P(\bar{B}|A)$ = probability that the disease is absent given that the factor is present

$$= b/(a + b).$$

Then the *odds* that a person with the factor present will have the disease is defined to be

$$O_A = P(B|A)/P(\bar{B}|A)$$
$$= a/b$$

Clearly, O_A lies between zero (no one with the factor gets the disease) and positive infinity (everyone with the factor gets the disease).

Table 9.2. Incidence of hypertension by oral contraceptive use

Using Contraceptives	Hypertensive Yes ($= B$)	No ($= \bar{B}$)	Total
Yes ($= A$)	a	b	$\underline{a + b}$
No ($= \bar{A}$)	c	d	$\underline{c + d}$
Total	$a + c$	$b + d$	$a + b + c + d$

Similarly, we define the odds for those without the factor to get the disease as

$$O_{\bar{A}} = P(B|\bar{A})/P(\bar{B}|\bar{A})$$
$$= c/d$$

where $O_{\bar{A}}$ also lies between zero and positive infinity.

Finally, we define the *odds ratio* as

$$o = O_A/O_{\bar{A}}$$
$$= P(B|A)P(\bar{B}|\bar{A})/P(B|\bar{A})P(\bar{B}|A)$$
$$= ad/bc$$

It will be noted that the odds ratio is formed from the cross products of the entries in the body of Table 9.1. For this reason it is sometimes called the *cross-product ratio*.

The odds ratio for a cross-sectional study is interpreted as follows: Subjects with the factor are *o times more likely to have the disease* than those without the factor. Note that the expected odds ratio is 1 if the disease and the factor are not associated with each other.

The *significance* of the association of the factor and the disease would be assessed by performing a chi-square analysis on the data in Table 9.1 (see Chapter 7). The strength of the association is measured by the odds ratio.

If the investigators had used a prospective study design, the data would appear as in Table 9.2, where $\underline{a + b}$ and $\underline{c + d}$ are underlined to indicate that $\underline{a + b}$ subjects free of the disease but exposed to the factor and $\underline{b + d}$ subjects free of the disease but not exposed to the factor were followed over time to measure the incidence of the disease in the two groups.

Clearly, $P(B|A)$, $P(\bar{B}|A)$, $P(B|\bar{A})$, and $P(\bar{B}|\bar{A})$ are defined as before and estimable from Table 9.2. Thus,

$$o = ad/bc$$

Table 9.3. Frequency of oral contraceptive use by hypertensive status

Use Oral Contraceptives	Hypertensive Yes	No	Total
Yes	a	b	$a + b$
No	c	d	$c + d$
Total	$\underline{a + c}$	$\underline{b + d}$	$a + b + c + d$

and the significance of the association between the disease and the factor would be calculated as before using chi-square.

However, the interpretation of the results in Table 9.2 differ from that for Table 9.1. In the cross-sectional study (Table 9.1) it is not known whether the (etiologic) factor of oral contraceptive use precedes the onset of the disease of hypertension. All that is known is whether the disease is *currently* more frequent among those who use oral contraceptives compared to those who do not. The odds ratio would be interpreted as "It is 'o' times more likely to find hypertension and oral contraceptive use occurring together in a subject than it is to find hypertension and no oral contraceptive use occurring together." The chi-square test would compare the proportion with the joint presence of the factor and disease to that for the disease without the factor.

In the prospective study (Table 9.2), it is known that the factor precedes the onset of the disease. Thus, the odds ratio would be interpreted as "It is 'o' times more likely that a subject who uses oral contraceptives will develop hypertension than it is for a subject who does not use oral contraceptives." The chi-square analysis will test the rate at which the disease develops in the two groups.

Had the researchers decided to use the retrospective approach to study the association of hypertension with oral contraceptive use, the data would appear as in Table 9.3. The totals $a + c$ and $b + d$ are underlined to indicate that the researchers located $a + c$ and $b + d$ subjects (or medical records), with and without documented hypertension respectively, and looked back in time to ascertain oral contraceptive use prior to the onset of the disease.

It is obvious that $P(B|A)$, $P(\bar{B}|A)$, $P(B|\bar{A})$, and $P(\bar{B}|\bar{A})$ cannot be defined from these data, since this would require the random selection of subjects both using and not using oral contraceptives, and this has not been done. Fortunately, we can use the following relationships involving joint probabilities (see Chapter 2):

$$P(B|A) = P(A \text{ and } B)/P(A)$$

$$= P(A|B)P(B)/P(A)$$

and

$$P(\bar{B}|A) = P(A \text{ and } \bar{B})/P(A)$$
$$= P(A|\bar{B})P(\bar{B})/P(A)$$

Thus, as defined before,

$$O_A = P(B|A)/P(\bar{B}|A)$$
$$= P(A|B)P(B)/P(A|\bar{B})P(\bar{B})$$

Similarly,

$$P(B|\bar{A}) = P(\bar{A} \text{ and } B)/P(\bar{A})$$
$$= P(\bar{A}|B)P(B)/P(\bar{A})$$

and

$$P(\bar{B}|\bar{A}) = P(\bar{A} \text{ and } \bar{B})/P(\bar{A})$$
$$= P(\bar{A}|\bar{B})P(\bar{B})/P(\bar{A})$$

Also, as before, we have,

$$O_{\bar{A}} = P(\bar{A}|B)P(B)/P(\bar{A}|\bar{B})P(\bar{B})$$

Thus,

$$o = O_A/O_{\bar{A}}$$
$$= P(A|B)P(\bar{A}/\bar{B})/P(A|\bar{B})P(\bar{A}|B)$$

where each term can be calculated from the data in Table 9.3.
In particular,

$$P(A|B) = a/(a + c)$$
$$P(\bar{A}|B) = c/(a + c)$$
$$P(A|\bar{B}) = b/(b + d)$$

and

$$P(\bar{A}|\bar{B}) = d/(b + d)$$

This yields

$$o = ad/bc$$

as before and will measure the relative likelihood that persons with the factor will develop the disease as in the cross-sectional and prospective studies. However, it is important to note that the chi-square analysis compares the rate at which the *factor* is found between the diseased and non-diseased groups. An excellent discussion of these issues is found in *Statistical Methods for Rates and Proportions* by J. S. Fleiss (New York: John Wiley and Sons, Inc., 1981). This text also should be consulted for computing standard errors and confidence intervals for odds ratios.

METHODOLOGY

Since the techniques for performing chi-square analyses have been covered in Chapter 7 and the computation of the odds ratio is similar for each type of study, examples of retrospective studies will be used to show numerical computations.

Independent samples. Suppose researchers reviewed the records of 100 patients with hypertension (cases) and of 100 patients without hypertension (controls) who were comparable in other respects to the cases. The history of the use of oral contraceptives was reviewed, and "used oral contraceptives" was appropriately defined. The data appear in Table 9.4.

The reader should show that the corrected chi-square is 4.27 with 1 df. Therefore we conclude that the use of oral contraceptives appears more often among the cases than among the controls.

Table 9.4. Frequency of oral contraceptive use in hypertensive and nonhypertensive patients

Use Oral Contraceptives	Hypertensive Yes	No	Total
Yes	28	15	43
No	72	82	157
Total	100	100	200

The odds ratio is

$$o = \frac{28 \times 85}{72 \times 15}$$

$$= 2.20$$

indicating that it is 2.20 times more likely to find hypertension among those who use oral contraceptives than among those who do not for the populations represented by this study.

Dependent samples.   For purposes of excluding the effects of other variables except the factor under study, matching of cases with controls is often carried to the point where each case is matched with its own particular control. As in the paired t-test (Chapter 4), the data are said to come from a *matched sample*, and the analysis must reflect the matching.

Suppose a group of investigators carefully matched each of 100 cases (patients with the disease) to an individual control (e.g., on age, sex, race, etc.) and then ascertained the presence or absence of the suspected factor. The data would be displayed as in Table 9.5.

Thus, among 13 pairs both the case and the matched control had the factor present, while among 17 pairs the case had the factor and the matched control did not.

Clearly, only cases and controls who differ on the presence of the factor give information about the differential rate of the factor among the populations represented by these cases and controls. The corrected chi-square test, known as *McNemar's Test*, is given by

$$\chi^2 = (|b - c| - 1)^2/(b + c) \qquad 1 \ df$$

The odds ratio is simply

$$o = b/c$$

Table 9.5.   Frequency of factor "x" among patients with disease "y" and their matched controls

Cases	Controls Factor Present	Factor Absent	Total
Factor present	13 ($= a$)	17 ($= b$)	30
Factor absent	7 ($= c$)	63 ($= d$)	70
Total	20	80	100

For the data in Table 9.5 we see that

$$\chi^2 = (|17 - 7| - 1)^2/(17 + 7)$$
$$= 3.38 \qquad 1 \; df$$

which is not significant at the $\alpha = .05$ level.

The odds ratio is

$$o = 17/7$$
$$= 2.43$$

which indicates that it is 2.43 times more likely to find the factor present among cases and absent in their controls than it is to find the factor absent in the cases but present in their controls. Although the odds ratio is greater than 1, the association is not statistically significant.

PROBLEMS

1. Using the data in Problem 3, Chapter 7, compute and interpret the odds ratio.
2. Using the data in Problem 4, Chapter 7, compute and interpret the odds ratio for stimulants causing 11 or more mistakes as compared to the placebo.
3. Using the data in Problem 1, Chapter 7, compute and interpret the odds ratio for obtaining a successful outcome nationally as compared to your hospital.

APPENDIX A

TABLES

Table A. Percentiles of the chi-square distribution

df	$\chi^2_{.005}$	$\chi^2_{.025}$	$\chi^2_{.05}$	$\chi^2_{.90}$	$\chi^2_{.95}$	$\chi^2_{.975}$	$\chi^2_{.99}$	$\chi^2_{.995}$
1	.0000393	.000982	.00393	2.706	3.841	5.024	6.635	7.879
2	.0100	.0506	.103	4.605	5.991	7.378	9.210	10.597
3	0.717	.216	.352	6.251	7.815	9.348	11.345	12.838
4	.207	.484	.711	7.779	9.488	11.143	13.277	14.860
5	.412	.831	1.145	9.236	11.070	12.832	15.086	16.750
6	.676	1.237	1.635	10.645	12.592	14.449	16.812	18.548
7	.989	1.690	2.167	12.017	14.067	16.013	18.475	20.278
8	1.344	2.180	2.733	13.362	15.507	17.535	20.090	21.955
9	1.735	2.700	3.325	14.684	16.919	19.023	21.666	23.589
10	2.156	3.247	3.940	15.987	18.307	20.483	23.209	25.188
11	2.603	3.816	4.575	17.275	19.675	21.920	24.725	26.757
12	3.074	4.404	5.226	18.549	21.026	23.336	26.217	28.300
13	3.565	5.009	5.892	19.812	22.362	24.736	27.688	29.819
14	4.075	5.629	6.571	21.064	23.685	26.119	29.141	31.319
15	4.601	6.262	7.261	22.307	24.996	27.488	30.578	32.801
16	5.142	6.908	7.962	23.542	26.296	28.845	32.000	34.267
17	5.697	7.564	8.672	24.769	27.587	30.191	33.409	35.718
18	6.265	8.231	9.390	25.989	28.869	31.526	34.805	37.156
19	6.844	8.907	10.117	27.204	30.144	32.852	36.191	38.582
20	7.434	9.591	10.851	28.412	31.410	34.170	37.566	39.997
21	8.034	10.283	11.591	29.615	32.671	35.479	38.932	41.401
22	8.643	10.982	12.338	30.813	33.924	36.781	40.289	42.796
23	9.260	11.688	13.091	32.007	35.172	38.076	41.638	44.181
24	9.886	12.401	13.848	33.196	36.415	39.364	42.980	45.558
25	10.520	13.120	14.611	34.382	37.652	40.646	44.314	46.928
26	11.160	13.844	15.379	35.563	38.885	41.923	45.642	48.290
27	11.808	14.573	16.151	36.741	40.113	43.194	46.963	49.645
28	12.461	15.308	16.928	37.916	41.337	44.461	48.278	50.993
29	13.121	16.047	17.708	39.087	42.557	45.722	49.588	52.336
30	13.787	16.791	18.493	40.256	43.773	46.979	50.892	53.672
35	17.192	20.569	22.465	46.059	49.802	53.203	57.342	60.275
40	20.707	24.433	26.509	51.805	55.758	59.342	63.691	66.766
45	24.311	28.366	30.612	57.505	61.656	65.410	69.957	73.166
50	27.991	32.357	34.764	63.167	67.505	71.420	76.154	79.490
60	35.535	40.482	43.188	74.397	79.082	83.298	88.379	91.952
70	43.275	48.758	51.739	85.527	90.531	95.023	100.425	104.215
80	51.172	57.153	60.391	96.578	101.879	106.629	112.329	116.321
90	59.196	65.647	69.126	107.565	113.145	118.136	124.116	128.299
100	67.328	74.222	77.929	118.498	124.342	129.561	135.807	140.169

SOURCE: A. Hald and S. A. Sinkbaek, "A Table of Percentage Points of the X Distribution," *Skandinavisk Aktuarietidskrift*, 33 (1950), 168–175. Published with permission from Skandinavisk Aktuarietidskrift.

Table B. Areas under the normal curve

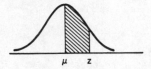

z	.00	.01	.02	.03	.04	.05	.06	.07	.08	.09
.0	.0000	.0040	.0080	.0120	.0160	.0199	.0239	.0279	.0319	.0359
.1	.0398	.0438	.0478	.0517	.0557	.0596	.0636	.0675	.0714	.0753
.2	.0793	.0832	.0871	.0910	.0948	.0987	.1026	.1064	.1103	.1141
.3	.1179	.1217	.1255	.1293	.1331	.1368	.1406	.1443	.1480	.1517
.4	.1554	.1591	.1628	.1664	.1700	.1736	.1772	.1808	.1844	.1879
.5	.1915	.1950	.1985	.2019	.2054	.2088	.2123	.2157	.2190	.2224
.6	.2257	.2291	.2324	.2357	.2389	.2422	.2454	.2486	.2517	.2549
.7	.2580	.2611	.2642	.2673	.2704	.2734	.2764	.2794	.2823	.2852
.8	.2881	.2910	.2939	.2967	.2995	.3023	.3051	.3078	.3106	.3133
.9	.3159	.3186	.3212	.3238	.3264	.3289	.3315	.3340	.3365	.3389
1.0	.3413	.3438	.3461	.3485	.3508	.3531	.3554	.3577	.3599	.3621
1.1	.3643	.3665	.3686	.3708	.3729	.3749	.3770	.3790	.3810	.3830
1.2	.3849	.3869	.3888	.3907	.3925	.3944	.3962	.3980	.3997	.4015
1.3	.4032	.4049	.4066	.4082	.4099	.4115	.4131	.4147	.4162	.4177
1.4	.4192	.4207	.4222	.4236	.4251	.4265	.4279	.4292	.4306	.4319
1.5	.4332	.4345	.4357	.4370	.4382	.4394	.4406	.4418	.4429	.4441
1.6	.4452	.4463	.4474	.4484	.4495	.4505	.4515	.4525	.4535	.4545
1.7	.4554	.4564	.4573	.4582	.4591	.4599	.4608	.4616	.4625	.4633
1.8	.4641	.4649	.4656	.4664	.4671	.4678	.4686	.4693	.4699	.4706
1.9	.4713	.4719	.4726	.4732	.4738	.4744	.4750	.4756	.4761	.4767
2.0	.4772	.4778	.4783	.4788	.4793	.4798	.4803	.4808	.4812	.4817
2.1	.4821	.4826	.4830	.4834	.4838	.4842	.4846	.4850	.4854	.4857
2.2	.4861	.4864	.4868	.4871	.4875	.4878	.4881	.4884	.4887	.4890
2.3	.4893	.4896	.4898	.4901	.4904	.4906	.4909	.4911	.4913	.4916
2.4	.4918	.4920	.4922	.4925	.4927	.4929	.4931	.4932	.4934	.4936
2.5	.4938	.4940	.4941	.4943	.4945	.4946	.4948	.4949	.4951	.4952
2.6	.4953	.4955	.4956	.4957	.4959	.4960	.4961	.4962	.4963	.4964
2.7	.4965	.4966	.4967	.4968	.4969	.4970	.4971	.4972	.4973	.4974
2.8	.4974	.4975	.4976	.4977	.4977	.4978	.4979	.4979	.4980	.4981
2.9	.4981	.4982	.4982	.4983	.4984	.4984	.4985	.4985	.4986	.4986
3.0	.4987	.4987	.4987	.4988	.4988	.4989	.4989	.4989	.4990	.4990

SOURCE: John E. Freund/Frank J. Williams, *Elementary Business Statistics: The Modern Approach*, 2nd ed., © 1972. Reprinted by permission of Prentice-Hall, Inc., Englewood Cliffs, N.J.

Table C. The t distribution

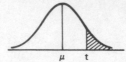

df	$t_{.10}$	$t_{.05}$	$t_{.025}$	$t_{.01}$	$t_{.005}$
1	3.078	6.3138	12.706	31.821	63.657
2	1.886	2.9200	4.3027	6.965	9.9248
3	1.638	2.3534	3.1825	4.541	5.8409
4	1.533	2.1318	2.7764	3.747	4.6041
5	1.476	2.0150	2.5706	3.365	4.0321
6	1.440	1.9432	2.4469	3.143	3.7074
7	1.415	1.8946	2.3646	2.998	3.4995
8	1.397	1.8595	2.3060	2.896	3.3554
9	1.383	1.8331	2.2622	2.821	3.2498
10	1.372	1.8125	2.2281	2.764	3.1693
11	1.363	1.7959	2.2010	2.718	3.1058
12	1.356	1.7823	2.1788	2.681	3.0545
13	1.350	1.7709	2.1604	2.650	3.0123
14	1.345	1.7613	2.1448	2.624	2.9768
15	1.341	1.7530	2.1315	2.602	2.9467
16	1.337	1.7459	2.1199	2.583	2.9208
17	1.333	1.7396	2.1098	2.567	2.8982
18	1.330	1.7341	2.1009	2.552	2.8784
19	1.328	1.7291	2.0930	2.539	2.8609
20	1.325	1.7247	2.0860	2.528	2.8453
21	1.323	1.7207	2.0796	2.518	2.8314
22	1.321	1.7171	2.0739	2.508	2.8188
23	1.319	1.7139	2.0687	2.500	2.8073
24	1.318	1.7109	2.0639	2.492	2.7969
25	1.316	1.7081	2.0595	2.485	2.7874
26	1.315	1.7056	2.0555	2.479	2.7787
27	1.314	1.7033	2.0518	2.473	2.7707
28	1.313	1.7011	2.0484	2.467	2.7633
29	1.311	1.6991	2.0452	2.462	2.7564
30	1.310	1.6973	2.0423	2.457	2.7500
35	1.3062	1.6896	2.0301	2.438	2.7239
40	1.3031	1.6839	2.0211	2.423	2.7045
45	1.3007	1.6794	2.0141	2.412	2.6896
50	1.2987	1.6759	2.0086	2.403	2.6778
60	1.2959	1.6707	2.0003	2.390	2.6603
70	1.2938	1.6669	1.9945	2.381	2.6480
80	1.2922	1.6641	1.9901	2.374	2.6388
90	1.2910	1.6620	1.9867	2.368	2.6316
100	1.2901	1.6602	1.9840	2.364	2.6260
120	1.2887	1.6577	1.9799	2.358	2.6175
140	1.2876	1.6558	1.9771	2.353	2.6114
160	1.2869	1.6545	1.9749	2.350	2.6070
180	1.2863	1.6534	1.9733	2.347	2.6035
200	1.2858	1.6525	1.9719	2.345	2.6006
∞	1.282	1.645	1.96	2.326	2.576

SOURCE: *Geigy Scientific Tables*, 7th edition. Basle, 1970.

Table D. Critical values of the correlation coefficient for different levels of significance

df*	.05	.01	df	.05	.01	df	.05	.01
1	.996917	.9998766	11	.5529	.6835	25	.3809	.4869
2	.95000	.990000	12	.5324	.6614	30	.3494	.4487
3	.8783	.95873	13	.5139	.6411	35	.3246	.4182
4	.8114	.91720	14	.4973	.6226	40	.3044	.3932
5	.7545	.8745	15	.4821	.6055	45	.2875	.3721
6	.7067	.8343	16	.4683	.5897	50	.2732	.3541
7	.6664	.7977	17	.4555	.5751	60	.2500	.3248
8	.6319	.7646	18	.4438	.5614	70	.2319	.3017
9	.6021	.7348	19	.4329	.5487	80	.2172	.2830
10	.5760	.7079	20	.4227	.5368	90	.2050	.2673
						100	.1946	.2540

* The degrees of freedom (*df*) are 2 less than the number of pairs in the sample.

SOURCE: Excerpted with permission of Hafner Press, a Division of Macmillan Publishing Co., Inc. from *Statistical Methods for Research Workers*, 14th Ed. Copyright © 1970 University of Adelaide.

Table E. F distribution ($\alpha = 0.05$)

Denominator Degrees of Freedom	Numerator Degrees of Freedom																		
	1	2	3	4	5	6	7	8	9	10	12	15	20	24	30	40	60	120	∞
1	161.4	199.5	215.7	224.6	230.2	234.0	236.8	238.9	240.5	241.9	243.9	245.9	248.0	249.1	250.1	251.1	252.2	253.3	254.3
2	18.51	19.00	19.16	19.25	19.30	19.33	19.35	19.37	19.38	19.40	19.41	19.43	19.45	19.45	19.46	19.47	19.48	19.49	19.50
3	10.13	9.55	9.28	9.12	9.01	8.94	8.89	8.85	8.81	8.79	8.74	8.70	8.66	8.64	8.62	8.59	8.57	8.55	8.53
4	7.71	6.94	6.59	6.39	6.26	6.16	6.09	6.04	6.00	5.96	5.91	5.86	5.80	5.77	5.75	5.72	5.69	5.66	5.63
5	6.61	5.79	5.41	5.19	5.05	4.95	4.88	4.82	4.77	4.74	4.68	4.62	4.56	4.53	4.50	4.46	4.43	4.40	4.36
6	5.99	5.14	4.76	4.53	4.39	4.28	4.21	4.15	4.10	4.06	4.00	3.94	3.87	3.84	3.81	3.77	3.74	3.70	3.67
7	5.59	4.74	4.35	4.12	3.97	3.87	3.79	3.73	3.68	3.64	3.57	3.51	3.44	3.41	3.38	3.34	3.30	3.27	3.23
8	5.32	4.46	4.07	3.84	3.69	3.58	3.50	3.44	3.39	3.35	3.28	3.22	3.15	3.12	3.08	3.04	3.01	2.97	2.93
9	5.12	4.26	3.86	3.63	3.48	3.37	3.29	3.23	3.18	3.14	3.07	3.01	2.94	2.90	2.86	2.83	2.79	2.75	2.74
10	4.96	4.10	3.71	3.48	3.33	3.22	3.14	3.07	3.02	2.98	2.91	2.85	2.77	2.74	2.70	2.66	2.62	2.58	2.54
11	4.84	3.98	3.59	3.36	3.20	3.09	3.01	2.95	2.90	2.85	2.79	2.72	2.65	2.61	2.57	2.53	2.49	2.45	2.40
12	4.75	3.89	3.49	3.26	3.11	3.00	2.91	2.85	2.80	2.75	2.69	2.62	2.54	2.51	2.47	2.43	2.38	2.34	2.30
13	4.67	3.81	3.41	3.18	3.03	2.92	2.83	2.77	2.71	2.67	2.60	2.53	2.46	2.42	2.38	2.34	2.30	2.25	2.21
14	4.60	3.74	3.34	3.11	2.96	2.85	2.76	2.70	2.65	2.60	2.53	2.46	2.39	2.35	2.31	2.27	2.22	2.18	2.13
15	4.54	3.68	3.29	3.06	2.90	2.79	2.71	2.64	2.59	2.54	2.48	2.40	2.33	2.29	2.25	2.20	2.16	2.11	2.07
16	4.49	3.63	3.24	3.01	2.85	2.74	2.66	2.59	2.54	2.49	2.42	2.35	2.28	2.24	2.19	2.15	2.11	2.06	2.01
17	4.45	3.59	3.20	2.96	2.81	2.70	2.61	2.55	2.49	2.45	2.38	2.31	2.23	2.19	2.15	2.10	2.06	2.01	1.96
18	4.41	3.55	3.16	2.93	2.77	2.66	2.58	2.51	2.46	2.41	2.34	2.27	2.19	2.15	2.11	2.06	2.02	1.97	1.92
19	4.38	3.52	3.13	2.90	2.74	2.63	2.54	2.48	2.42	2.38	2.31	2.23	2.16	2.11	2.07	2.03	1.98	1.93	1.88
20	4.35	3.49	3.10	2.87	2.71	2.60	2.51	2.45	2.39	2.35	2.28	2.20	2.12	2.08	2.04	1.99	1.95	1.90	1.84
21	4.32	3.47	3.07	2.84	2.68	2.57	2.49	2.42	2.37	2.32	2.25	2.18	2.10	2.05	2.01	1.96	1.92	1.87	1.81
22	4.30	3.44	3.05	2.82	2.66	2.55	2.46	2.40	2.34	2.30	2.23	2.15	2.07	2.03	1.98	1.94	1.89	1.84	1.78
23	4.28	3.42	3.03	2.80	2.64	2.53	2.44	2.37	2.32	2.27	2.20	2.13	2.05	2.01	1.96	1.91	1.86	1.81	1.76
24	4.26	3.40	3.01	2.78	2.62	2.51	2.42	2.36	2.30	2.25	2.18	2.11	2.03	1.98	1.94	1.89	1.84	1.79	1.73
25	4.24	3.39	2.99	2.76	2.60	2.49	2.40	2.34	2.28	2.24	2.16	2.09	2.01	1.96	1.92	1.87	1.82	1.77	1.71
26	4.23	3.37	2.98	2.74	2.59	2.47	2.39	2.32	2.27	2.22	2.15	2.07	1.99	1.95	1.90	1.85	1.80	1.75	1.69
27	4.21	3.35	2.96	2.73	2.57	2.46	2.37	2.31	2.25	2.20	2.13	2.06	1.97	1.93	1.88	1.84	1.79	1.73	1.67
28	4.20	3.34	2.95	2.71	2.56	2.45	2.36	2.29	2.24	2.19	2.12	2.04	1.96	1.91	1.87	1.82	1.77	1.71	1.65
29	4.18	3.33	2.93	2.70	2.55	2.43	2.35	2.28	2.22	2.18	2.10	2.03	1.94	1.90	1.85	1.81	1.75	1.70	1.64
30	4.17	3.32	2.92	2.69	2.53	2.42	2.33	2.27	2.21	2.16	2.09	2.01	1.93	1.89	1.84	1.79	1.74	1.68	1.62
40	4.08	3.23	2.84	2.61	2.45	2.34	2.25	2.18	2.12	2.08	2.00	1.92	1.84	1.79	1.74	1.69	1.64	1.58	1.51
60	4.00	3.15	2.76	2.53	2.37	2.25	2.17	2.10	2.04	1.99	1.92	1.84	1.75	1.70	1.65	1.59	1.53	1.47	1.39
120	3.92	3.07	2.68	2.45	2.29	2.17	2.09	2.02	1.96	1.91	1.83	1.75	1.66	1.61	1.55	1.50	1.43	1.35	1.25
∞	3.84	3.00	2.60	2.37	2.21	2.10	2.01	1.94	1.88	1.83	1.75	1.67	1.57	1.52	1.46	1.39	1.32	1.22	1.00

SOURCE: *Biometrika Tables for Statisticians*, Third Edition, Vol. I, London, 1966. By permission of Biometrika Trustees.

Table E. F distribution ($\alpha = 0.01$)

Denominator Degrees of Freedom	\	\	\	\	Numerator Degrees of Freedom	\	\	\	\	\	\	\	\	\	\	\	\	\	\
	1	2	3	4	5	6	7	8	9	10	12	15	20	24	30	40	60	120	∞
1	4052	4999.5	5403	5625	5764	5859	5928	5981	6022	6056	6106	6157	6209	6235	6261	6287	6313	6339	6366
2	98.50	99.00	99.17	99.25	99.30	99.33	99.36	99.37	99.39	99.40	99.42	99.43	99.45	99.46	99.47	99.47	99.48	99.49	99.50
3	34.12	30.82	29.46	28.71	28.24	27.91	27.67	27.49	27.35	27.23	27.05	26.87	26.69	26.60	26.50	26.41	26.32	26.22	26.13
4	21.20	18.00	16.69	15.98	15.52	15.21	14.98	14.80	14.66	14.55	14.37	14.20	14.02	13.93	13.84	13.75	13.65	13.56	13.46
5	16.26	13.27	12.06	11.39	10.97	10.67	10.46	10.29	10.16	10.05	9.89	9.72	9.55	9.47	9.38	9.29	9.20	9.11	9.02
6	13.75	10.92	9.78	9.15	8.75	8.47	8.26	8.10	7.98	7.87	7.72	7.56	7.40	7.31	7.23	7.14	7.06	6.97	6.88
7	12.25	9.55	8.45	7.85	7.46	7.19	6.99	6.84	6.72	6.62	6.47	6.31	6.16	6.07	5.99	5.91	5.82	5.74	5.65
8	11.26	8.65	7.59	7.01	6.63	6.37	6.18	6.03	5.91	5.81	5.67	5.52	5.36	5.28	5.20	5.12	5.03	4.95	4.86
9	10.56	8.02	6.99	6.42	6.06	5.80	5.61	5.47	5.35	5.26	5.11	4.96	4.81	4.73	4.65	4.57	4.48	4.40	4.31
10	10.04	7.56	6.55	5.99	5.64	5.39	5.20	5.06	4.94	4.85	4.71	4.56	4.41	4.33	4.25	4.17	4.08	4.00	3.91
11	9.65	7.21	6.22	5.67	5.32	5.07	4.89	4.74	4.63	4.54	4.40	4.25	4.10	4.02	3.94	3.86	3.78	3.69	3.60
12	9.33	6.93	5.95	5.41	5.06	4.82	4.64	4.50	4.39	4.30	4.16	4.01	3.86	3.78	3.70	3.62	3.54	3.45	3.36
13	9.07	6.70	5.74	5.21	4.86	4.62	4.44	4.30	4.19	4.10	3.96	3.82	3.66	3.59	3.51	3.43	3.34	3.25	3.17
14	8.86	6.51	5.56	5.04	4.69	4.46	4.28	4.14	4.03	3.94	3.80	3.66	3.51	3.43	3.35	3.27	3.18	3.09	3.00
15	8.68	6.36	5.42	4.89	4.56	4.32	4.14	4.00	3.89	3.80	3.67	3.52	3.37	3.29	3.21	3.13	3.05	2.96	2.87
16	8.53	6.23	5.29	4.77	4.44	4.20	4.03	3.89	3.78	3.69	3.55	3.41	3.26	3.18	3.10	3.02	2.93	2.84	2.75
17	8.40	6.11	5.18	4.67	4.34	4.10	3.93	3.79	3.68	3.59	3.46	3.31	3.16	3.08	3.00	2.92	2.83	2.75	2.65
18	8.29	6.01	5.09	4.58	4.25	4.01	3.84	3.71	3.60	3.51	3.37	3.23	3.08	3.00	2.92	2.84	2.75	2.66	2.57
19	8.18	5.93	5.01	4.50	4.17	3.94	3.77	3.63	3.52	3.43	3.30	3.15	3.00	2.92	2.84	2.76	2.67	2.58	2.49
20	8.10	5.85	4.94	4.43	4.10	3.87	3.70	3.56	3.46	3.37	3.23	3.09	2.94	2.86	2.78	2.69	2.61	2.52	2.42
21	8.02	5.78	4.87	4.37	4.04	3.81	3.64	3.51	3.40	3.31	3.17	3.03	2.88	2.80	2.72	2.64	2.55	2.46	2.36
22	7.95	5.72	4.82	4.31	3.99	3.76	3.59	3.45	3.35	3.26	3.12	2.98	2.83	2.75	2.67	2.58	2.50	2.40	2.31
23	7.88	5.66	4.76	4.26	3.94	3.71	3.54	3.41	3.30	3.21	3.07	2.93	2.78	2.70	2.62	2.54	2.45	2.35	2.26
24	7.82	5.61	4.72	4.22	3.90	3.67	3.50	3.36	3.26	3.17	3.03	2.89	2.74	2.66	2.58	2.49	2.40	2.31	2.21
25	7.77	5.57	4.68	4.18	3.85	3.63	3.46	3.32	3.22	3.13	2.99	2.85	2.70	2.62	2.54	2.45	2.36	2.27	2.17
26	7.72	5.53	4.64	4.14	3.82	3.59	3.42	3.29	3.18	3.09	2.96	2.81	2.66	2.58	2.50	2.42	2.33	2.23	2.13
27	7.68	5.49	4.60	4.11	3.78	3.56	3.39	3.26	3.15	3.06	2.93	2.78	2.63	2.55	2.47	2.38	2.29	2.20	2.10
28	7.64	5.45	4.57	4.07	3.75	3.53	3.36	3.23	3.12	3.03	2.90	2.75	2.60	2.52	2.44	2.35	2.26	2.17	2.06
29	7.60	5.42	4.54	4.04	3.73	3.50	3.33	3.20	3.09	3.00	2.87	2.73	2.57	2.49	2.41	2.33	2.23	2.14	2.03
30	7.56	5.39	4.51	4.02	3.70	3.47	3.30	3.17	3.07	2.98	2.84	2.70	2.55	2.47	2.39	2.30	2.21	2.11	2.01
40	7.31	5.18	4.31	3.83	3.51	3.29	3.12	2.99	2.89	2.80	2.66	2.52	2.37	2.29	2.20	2.11	2.02	1.92	1.80
60	7.08	4.98	4.13	3.65	3.34	3.12	2.95	2.82	2.72	2.63	2.50	2.35	2.20	2.12	2.03	1.94	1.84	1.73	1.60
120	6.85	4.79	3.95	3.48	3.17	2.96	2.79	2.66	2.56	2.47	2.34	2.19	2.03	1.95	1.86	1.76	1.66	1.53	1.38
∞	6.63	4.61	3.78	3.32	3.02	2.80	2.64	2.51	2.41	2.32	2.18	2.04	1.88	1.79	1.70	1.59	1.47	1.32	1.00

SOURCE: *Biometrika Tables for Statisticians*, Third Edition, Vol. I, London, 1966. By permission of Biometrika Trustees.

Table F. Student's *t* distribution: Number of observations for *t* test of mean

Level of t-Test

Single-Sided Test →	α = 0.005					α = 0.01					α = 0.025					α = 0.05				
Double-Sided Test →	α = 0.01					α = 0.02					α = 0.05					α = 0.1				
β =	0.01	0.05	0.1	0.2	0.5	0.01	0.05	0.1	0.2	0.5	0.01	0.05	0.1	0.2	0.5	0.01	0.05	0.1	0.2	0.5
$\Delta = \dfrac{\mu - \mu_0}{\sigma}$																				
0.05																				
0.10																				
0.15																				122
0.20										139					99				139	70
0.25					110					90				128	64			139	101	45
0.30				134	78				115	63			119	90	45		122	97	71	32
0.35			125	99	58			109	85	47		109	88	67	34		90	72	52	24
0.40		115	97	77	45		101	85	66	37	117	84	68	51	26	101	70	55	40	19
0.45		92	77	62	37	110	81	68	53	30	93	68	54	41	21	80	55	44	33	15
0.50	100	75	63	51	30	90	66	55	43	25	76	54	44	34	18	65	45	36	27	13
0.55	83	63	53	42	26	75	55	46	36	21	63	45	37	28	15	54	38	30	22	11
0.60	71	53	45	36	22	63	47	39	31	18	53	38	32	24	13	46	32	26	19	9
0.65	61	46	39	31	20	55	41	34	27	16	46	33	27	21	12	39	28	22	17	8
0.70	53	40	34	28	17	47	35	30	24	14	40	29	24	19	10	34	24	19	15	8
0.75	47	36	30	25	16	42	31	27	21	13	35	26	21	16	9	30	21	17	13	7
0.80	41	32	27	22	14	37	28	24	19	12	31	24	19	15	9	27	19	15	12	6
0.85	37	29	24	20	13	33	25	21	17	11	28	22	17	13	8	24	17	14	11	6
0.90	34	26	22	18	12	29	23	19	16	10	25	21	16	12	7	21	15	13	10	5

Value of $\Delta=\frac{\mu-\mu_0}{\sigma}$ · 0.95	31	24	20	17	11	27	21	18	14	9	23	17	14	11	7	19	14	11	9	5
1.00	28	22	19	16	10	25	19	16	13	9	21	16	13	10	6	18	13	11	8	5
1.1	24	19	16	14	9	21	16	14	12	8	18	13	11	9	6	15	11	9	7	5
1.2	21	16	14	12	8	18	14	12	10	7	15	12	10	8	5	13	10	8	6	
1.3	18	15	13	11	8	16	13	11	9	6	14	10	9	7		11	8	7	6	
1.4	16	13	12	10	7	14	11	10	9	6	12	9	8	7		10	8	7	5	
1.5	15	12	11	9	7	13	10	9	8	6	11	8	7	6		9	7	6		
1.6	13	11	10	8	6	12	10	9	7	5	10	8	7	6		8	6	6		
1.7	12	10	9	8	6	11	9	8	7		9	7	6	5		8	6	5		
1.8	12	10	9	8	6	10	8	7	7		8	7	6			7	6			
1.9	11	9	8	7	6	10	8	7	6		8	6	6			7	5			
2.0	10	8	8	7	5	9	7	7	6		7	6	5			6				
2.1	10	8	7	7		8	7	6	6		7	6								
2.2	9	8	7	6		8	7	6	5		7	6								
2.3	9	7	7	6		8	6	6			6	5								
2.4	8	7	7	6		7	6	6			6									
2.5	8	7	6	6		7	6	6			6									
3.0	7	6	6	5		6	5	5			5									
3.5	6	5	5			5														
4.0	6																			

SOURCE: Reprinted with permission from *CRC Handbook of Tables for Probability and Statistics*, 2nd ed., 1968. Copyright The Chemical Rubber Co., CRC Press, Inc.

Table F. Student's t distribution: (continued): Number of observations for t test of difference between two means

Level of t-Test

| Single-Sided Test → | α = 0.005 | | | | | α = 0.01 | | | | | α = 0.025 | | | | | α = 0.05 | | | | |
| Double-Sided Test → | α = 0.01 | | | | | α = 0.02 | | | | | α = 0.05 | | | | | α = 0.1 | | | | |
$\Delta = \dfrac{\mu_x - \mu_y}{\sigma}$ ＼ β =	0.01	0.05	0.1	0.2	0.5	0.01	0.05	0.1	0.2	0.5	0.01	0.05	0.1	0.2	0.5	0.01	0.05	0.1	0.2	0.5
0.05																				
0.10																				
0.15																				
0.20																				137
0.25															124					88
0.30										123					87					61
0.35					110					90					64				102	45
0.40					85					70				100	50			108	78	35
0.45				118	68					55			105	79	39		108	86	62	28
0.50				96	55		106	106	82	45		106	86	64	32		88	70	51	23
0.55			101	79	46		106	88	68	38		87	71	53	27	112	73	58	42	19
0.60		101	85	67	39		90	74	58	32	104	74	60	45	23	89	61	49	36	16
0.65		87	73	57	34	104	77	64	49	27	88	63	51	39	20	76	52	42	30	14
0.70	100	75	63	50	29	90	66	55	43	24	76	55	44	34	17	66	45	36	26	12
0.75	88	66	55	44	26	79	58	48	38	21	67	48	39	29	15	57	40	32	23	11
0.80	77	58	49	39	23	70	51	43	33	19	59	42	34	26	14	50	35	28	21	10
0.85	69	51	43	35	21	62	46	38	30	17	52	37	31	23	12	45	31	25	18	9
0.90	62	46	39	31	19	55	41	34	27	15	47	34	27	21	11	40	28	22	16	8

204

Value of $\Delta = \dfrac{\mu_x - \mu_1}{\sigma}$	55 50	42 38	35 32	28 26	17 15	50 45	37 33	31 28	24 22	14 13	42 38	30 27	25 23	19 17	10 9	36 33	25 23	20 18	15 14	7 7	Value of $\Delta = \dfrac{\mu_x - \mu_1}{\sigma}$
0.95 / 1.00	55 / 50	42 / 38	35 / 32	28 / 26	17 / 15	50 / 45	37 / 33	31 / 28	24 / 22	14 / 13	42 / 38	30 / 27	25 / 23	19 / 17	10 / 9	36 / 33	25 / 23	20 / 18	15 / 14	7 / 7	0.95 / 1.00
1.1	42	32	27	22	13	38	28	23	19	11	32	23	19	14	8	27	19	15	12	6	1.1
1.2	36	27	23	18	11	32	24	20	16	9	27	20	16	12	7	23	16	13	10	5	1.2
1.3	31	23	20	16	10	28	21	17	14	8	23	17	14	11	6	20	14	11	9	5	1.3
1.4	27	20	17	14	9	24	18	15	12	8	20	15	12	10	6	17	12	10	8	4	1.4
1.5	24	18	15	13	8	21	16	14	11	7	18	13	11	9	5	15	11	9	7	4	1.5
1.6	21	16	14	11	7	19	14	12	10	6	16	12	10	8	5	14	10	8	6	4	1.6
1.7	19	15	13	10	7	17	13	11	9	6	14	11	9	7	4	12	9	7	6	3	1.7
1.8	17	13	11	10	6	15	12	10	8	5	13	10	9	6	4	11	8	7	5		1.8
1.9	16	12	11	9	6	14	11	9	8	5	12	9	8	6	4	10	7	6	5		1.9
2.0	14	11	10	8	6	13	10	9	7	5	11	8	7	6	4	9	7	6	4		2.0
2.1	13	10	9	8	5	12	9	8	7	5	10	8	7	5	3	8	6	5	4		2.1
2.2	12	10	8	7	5	11	9	7	6	4	9	7	6	5		8	6	5	4		2.2
2.3	11	9	8	7	5	10	8	7	6	4	9	7	6	5		7	5	5	4		2.3
2.4	11	9	8	6	5	10	8	6	6	4	8	6	5	4		7	5	4	4		2.4
2.5	10	8	7	6	4	9	7	6	5	4	8	6	5	4		6	5	4	3		2.5
3.0	8	6	6	5	4	7	6	5	4	3	6	5	4	4		5	4	3			3.0
3.5	6	5	5	4	3	6	5	4	4		5	4	4	3		4	3				3.5
4.0	6	5	4	4		5	4	4	3		5	4	3			4					4.0

SOURCE: Reprinted with permission from *CRC Handbook of Tables for Probability and Statistics*, 2nd ed., 1968. Copyright The Chemical Rubber Co., CRC Press, Inc.

APPENDIX B

SOLUTIONS TO PROBLEMS

CHAPTER 1 SOLUTIONS

1. a. Frequency table for serum cholesterol values in 30-year-old males

Cholesterol Level (mg/100 ml)	Frequency	Relative Frequency $\frac{F}{n}$	Cumulative Frequency
119.5–139.5	1	.033	.033
139.5–159.5	2	.067	.100
159.5–179.5	3	.100	.200
179.5–199.5	4	.133	.333
199.5–219.5	7	.233	.566
219.5–239.5	6	.200	.766
239.5–259.5	4	.133	.899
259.5–279.5	2	.067	.966
279.5–299.5	1	.033	.999

b.

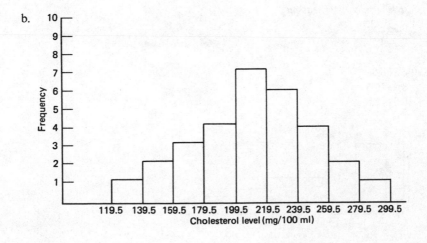

c.   i. $\bar{Y} = \dfrac{\sum Y}{n} = \dfrac{6311}{30} = 210.37$

    ii. Median $= \dfrac{210 + 215}{2} = 212.5$

    iii. Mode $= 200, 230$ (bimodal)

    iv. Range $= 290 - 120 = 170$

    v. $s^2 = \dfrac{\sum Y^2 - \dfrac{(\sum Y)^2}{n}}{n-1} = \dfrac{1371111 - \dfrac{(6311)^2}{30}}{29} = 1499.55$

    vi. $s = \sqrt{1499.55} = 38.72$

2.

Y_i	$(Y_i - \bar{Y})$	$(Y_i - \bar{Y})^2$	Y_i^2
5	$+2$	4	25
3	$+0$	0	9
2	-1	1	4
1	-2	4	1
4	$+1$	1	16
Totals 15	0	10	55

$$s^2 = \frac{\sum(Y_i - \bar{Y})^2}{n-1} = \frac{10}{4} = 2.5$$

$$s^2 = \frac{\sum Y^2 - \dfrac{(\sum Y)^2}{n}}{n-1} = \frac{55 - \dfrac{(15)^2}{5}}{4} = \frac{10}{4} = 2.5$$

3. *Class Intervals*   *Frequency*

     0.5–2.5   6

     2.5–4.5   5

     4.5–6.5   3

     6.5–8.5   3

     8.5–10.5   2

     10.5–12.5   1

a.

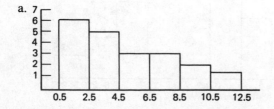

b. Skewed

c. $\bar{Y} = 4.8$ Median = 4.0

4. Phenacetin ($\mu g/ml$)	Nonsmokers Relative Frequency	Cumulative Frequency	Smokers Relative Frequency	Cumulative Frequency
0.005–0.505	.083	.083	.400	.400
0.505–1.005	.083	.166	.200	.600
1.005–1.505	.167	.333	.200	.800
1.505–2.005	.250	.583	.100	.900
2.005–2.505	.083	.666	0	.900
2.505–3.005	.167	.833	.100	1.00
3.005–3.505	.083	.916	0	1.00
3.505–4.005	.083	.999	0	1.00

b. Relative frequency is more meaningful since the totals of individuals in the two groups are not equal.

c. 58.3% 40%

d. Skewed

e. The median will be larger than the mean for the smoking group.

f. Based on the frequency tables, it appears that smokers have lower phenacetin concentrations than nonsmokers.

5. a.

Instrument	Mean	Standard Deviation	b.
1	10	4.055	Unbiased, not precise
2	9	1.155	Biased, precise
3	10	1.155	Unbiased, precise

c. Instrument 3 is the most accurate since it is both unbiased and precise.

6. a.

Lab	Technician	Mean	Standard Deviation	Variance
1	1	6.33	1.53	2.34
	2	5.33	2.31	5.34
2	1	6.67	2.31	5.34
	2	5.67	0.58	0.34

b. Technician 2 in lab 2 probably produces the most accurate results.

c.

Lab	Mean	Standard Deviation	Variance
1	5.83	0.707	0.5
2	6.17	0.707	0.5

d. Different technicians, different equipment.

e. Lab 1 appears to produce the most accurate results.

CHAPTER 2 SOLUTIONS

1. 12 students: 4 getting the new product, 4 getting the standard product, and 4 getting the placebo.

   a. $P(\text{new product}) = \frac{4}{12} = \frac{1}{3}$

   b. $P(\text{chemically active}) = P(\text{new product}) + P(\text{standard product})$
   $$= \frac{1}{3} + \frac{1}{3} = \frac{2}{3}$$

   $P(\text{chemically active}) = 1 - P(\text{placebo})$
   $$= 1 - \frac{1}{3}$$
   $$= \frac{2}{3}$$

   c. No

2. Chance of usable records $= 28/100$
   Chance of unusable records $= 72/100$
   $.28x = 100$
   $x = 358$ records necessary
   To check 358 records, expect to spend (2 min/record) $\times$ (358 records) $= 716$ minutes.

3. Let X = probability of failure for any given alarm. We wish to determine the probability that *both* alarms fail to signal.

   Possible outcomes:

Alarm 1	Alarm 2
Signal (S)	Not signal (NS)
Not signal	Signal
Signal	Signal
* Not signal	Not signal

   Using binomial notation,

   $n = 2$ p = probability that any one machine will fail
   $= .001$ ✓

   $P(X = 2) = \text{probability both fail} = \binom{2}{2}(.001)^2(.999)^0 = (.001)^2$ ✓

   Thus, there is one chance in a million that a cardiac arrest will not be signalled. Alternately, probability of failure $= 1 - P(\text{success})$ where success is the event that at least one of the machines will signal. There are three possible ways a "success" can occur. Therefore,

   $P(\text{failure}) = 1 - P(\text{success})$
   $1 - [(.999)(.001) + (.001)(.999) + (.999)(.999)]$
   $1 - [(NS)(S) + (S)(NS) + (S)(S)]$

In terms of the binomial formula

$$1 - P[X = 1 \text{ or } X = 2]$$

where X is a "success."

$$1 - \left[\binom{2}{1}(.999)^1(.001)^1 + \binom{2}{2}(.999)^2(.001)^0 \right]$$

4.

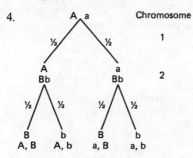

Probability of A, b = $\frac{1}{2} \times \frac{1}{2} = \frac{1}{4}$

5. Let X = the number of boys.

The possible outcomes are

* b b b b	g g b b
b b b g	b g g b
b b g b	g b g b
b g b b	b g g g
g b b b	g b g g
b b g g	g g b g
b g b g	g g g b
g b b g	g g g g

16   possible outcomes

a. There is only 1 out of 16 ways that all four births will be boys. Thus

$$P(bbbb) = \tfrac{1}{16}$$

This may be found without enumerating the sample space.

$$P(4 \text{ boys}) = 1(\tfrac{1}{2} \cdot \tfrac{1}{2} \cdot \tfrac{1}{2} \cdot \tfrac{1}{2}) = \tfrac{1}{16}$$

Using binomial notation where $n = 4$, $p = .5$, X = number of boys

$$P(X = 4) = \binom{4}{4}(.5)^4(.5)^0 = \tfrac{1}{16} \; \checkmark$$

b. Returning to the sample space, we see that there are only 2 ways that 2 of each sex are born alternately (i.e., bgbg, gbgb). Thus

$$P(\text{bgbg or gbgb}) = (\tfrac{1}{2} \cdot \tfrac{1}{2} \cdot \tfrac{1}{2} \cdot \tfrac{1}{2}) + (\tfrac{1}{2} \cdot \tfrac{1}{2} \cdot \tfrac{1}{2} \cdot \tfrac{1}{2})$$
$$= \tfrac{1}{16} + \tfrac{1}{16} = \tfrac{1}{8}$$

6. a. Father → A, B, C   Mother → a, b, c

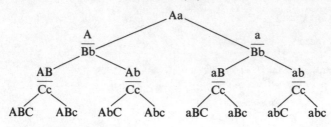

b. $P(\text{Abc}) = \frac{1}{8} = \frac{1}{2} \cdot \frac{1}{2} \cdot \frac{1}{2}$

$P(\text{ABC or abc}) = \frac{1}{8} + \frac{1}{8} = \frac{1}{4}$

7. Let A = specific allele

Aa man

child A, $\frac{1}{2}$ chance

grandchild A, $\frac{1}{2}$ chance

Therefore, the probability of passing a specific allele to a grandchild is

$(\frac{1}{2})(\frac{1}{2}) = \frac{1}{4}$

8. Desirable trait: Aa Bb

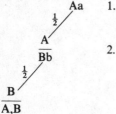

1. There is a $\frac{1}{2} \times \frac{1}{2} = \frac{1}{4}$ chance of passing A,B from a parent to a child.

2.

Thus, there is a $\frac{1}{4}$ chance that grandfather passes the genes to his child. The probability that this child passes the genes to his child is also $\frac{1}{4}$. Then the probability that the grandparent will pass these genes to a grandchild is $\frac{1}{4} \times \frac{1}{4} = \frac{1}{16}$.

9. Let X = the number of people who do not have the condition.

p = probability of "success"

  = probability that a person does not have the condition

  = $1 - .80 = .2$

$n = 5$

Therefore,

$P(X = 5)$ = probability that 5 people do not have the condition

$P(X = 5) = \binom{5}{5}(.2)^5(.8)^0 = (.2)^5$

10. a. Let I = ill $\bar{I}$ = not ill

       G = ate gravy $\bar{G}$ = no gravy

$$P(I|G) = \frac{P(I \text{ and } G)}{P(G)}$$

$$P(I \text{ and } G) = \frac{75}{115}$$

$$P(G) = \frac{100}{115}$$

Thus,

$$P(I|G) = \frac{\dfrac{75}{115}}{\dfrac{100}{115}} = \frac{75}{100} = .75$$

b. $P(I|\bar{G}) = \dfrac{\dfrac{5}{115}}{\dfrac{15}{115}} = \dfrac{5}{15} = \dfrac{1}{3} = .33$

c. $P(\bar{I}|G) = \dfrac{\dfrac{25}{115}}{\dfrac{100}{115}} = \dfrac{25}{100} = .25$

11. P = positive test

     D = disease

$$P(P|D) = .93$$
$$P(P|\bar{D}) = .03$$
$$P(D) = .01$$
$$P(\bar{D}) = .99$$

$$P(D|P) = \frac{P(P|D)P(D)}{P(P|D)P(D) + P(P|\bar{D})P(\bar{D})}$$

$$= \frac{(.93)(.01)}{(.93)(.01) + (.03)(.99)}$$

$$= .24$$

Prevalence = .01

Sensitivity = $P(P|D)$ = .93

Specificity = $1 - P(P|\bar{D}) = 1 - .03 = .97$

Predictive ability = $P(D|P)$ = .24

12. P = positive test

$\bar{P}$ = negative test

D = disease

$\bar{D}$ = no disease

$P(P|\bar{D}) = .07$

$P(\bar{P}|D) = .18$

$P(P|D) = .82$

$P(\bar{P}|\bar{D}) = .93$

$P(D) = .03$

a.

$$P(D|P) = \frac{P(P|D)P(D)}{P(P|D)P(D) + P(P|\bar{D})P(\bar{D})}$$

$$= \frac{(.82)(.03)}{(.82)(.03) + (.07)(.97)}$$

$$= .27$$

b.

$$P(D|\bar{P}) = \frac{P(\bar{P}|D)P(D)}{P(\bar{P}|D)P(D) + P(\bar{P}|\bar{D})P(\bar{D})}$$

$$= \frac{(.18)(.03)}{(.18)(.03) + (.93)(.97)}$$

$$= .006$$

Prevalence $= .03$

Sensitivity $= 1 - P(\bar{P}|D) = 1 - .18 = .82$

Specificity $= 1 - P(P|\bar{D}) = 1 - .03 = .97$

Predictive ability $= P(D|P) = .27$

CHAPTER 3 SOLUTIONS

1. a. $P(219.5 \leq Y \leq 259.5) = P\left[\dfrac{219.5 - 242.2}{45.4} \leq Z \leq \dfrac{259.5 - 242.2}{45.4}\right]$

$P(-0.5 \leq Z \leq 0.38) = .1915 + .1480 = .3395$

b. $P(139.5 \leq Y \leq 219.5) = P(-2.26 \leq Z \leq -0.5) = .4881 - .1915 = .2966$

c. $P(159.5 \leq Y \leq 179.5) = P(-1.82 \leq Z \leq -1.38) = .4656 - .4162 = .0494$

d. $P(Y \geq 242.2) = P(Z \geq 0) = 0.5$

e. $P(Y \leq 219.5) = P(Z \leq -0.5) = .5 - .1915 = .3085$

2. $P(-1.96 \leq Z \leq 1.96) = .4750 + .4750 = 0.95$

   $P(-2.58 \leq Z \leq 2.58) = .4950 + .4950 = 0.99$

3. a.

Plasma Potassium Level mEq/liter	Frequency	Relative Frequency	Theoretical Frequency
2.35–2.55	1	.033	.032
2.55–2.75	3	.100	.060
2.75–2.95	2	.067	.097
2.95–3.15	4	.133	.133
3.15–3.35	5	.167	.155
3.35–3.55	6	.200	.155
3.55–3.75	3	.100	.133
3.75–3.95	3	.100	.097
3.95–4.15	2	.067	.060
4.15–4.35	1	.033	.032

$P(2.35 \leq Y \leq 2.55) = P(-2.0 \leq Z \leq -1.6) = .4772 - .4452 = .032$

$P(2.55 \leq Y \leq 2.75) = P(-1.6 \leq Z \leq -1.2) = .4452 - .3849 = .060$

$P(2.75 \leq Y \leq 2.95) = P(-1.2 \leq Z \leq -0.8) = .3849 - .2881 = .097$

$P(2.95 \leq Y \leq 3.15) = P(-0.8 \leq Z \leq -0.4) = .2881 - .1554 = .133$

$P(3.15 \leq Y \leq 3.35) = P(-0.4 \leq Z \leq 0) = .1554$

$P(3.35 \leq Y \leq 3.55) = P(0 \leq Z \leq 0.4) = .1554$

$P(3.55 \leq Y \leq 3.75) = P(0.4 \leq Z \leq 0.8) = .2881 - .1554 = .133$

$P(3.75 \leq Y \leq 3.95) = P(0.8 \leq Z \leq 1.2) = .3849 - .2881 = .097$

$P(3.95 \leq Y \leq 4.15) = P(1.2 \leq Z \leq 1.6) = .4452 - .3849 = .060$

$P(4.15 \leq Y \leq 4.35) = P(1.6 \leq Z \leq 2.0) = .4772 - .4452 = .032$

b. The validity of the assumption of normality for the underlying theoretical distribution would be in doubt.

4. a. $P(Y \leq 150) = P(Z \leq 1.5) = 0.5 + 0.4332 = 0.9332$

   $P(Y \leq 110) = P(Z \leq -0.5) = 0.5 - 0.1915 = 0.3085$

   b. $Z = \dfrac{Y - \mu}{\sigma} = \dfrac{160 - 120}{20} = 2$

   c. $Z = \dfrac{Y - \mu}{\sigma}$

   $1.96 = \dfrac{Y - 120}{20}$ $\qquad$ $-1.96 = \dfrac{Y - 120}{20}$

   $Y = 159.2$ $\qquad\qquad$ $Y = 80.8$ $\qquad$ $P(80.8 \leq Y \leq 159.2) = .95$

   90% of the observations lie within ± 1.65 standard deviations.

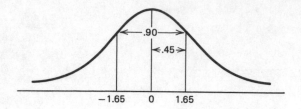

$$+1.65 = \frac{Y - 120}{20} \qquad -1.65 = \frac{Y - 120}{20}$$

$$Y = 153 \qquad\qquad Y = 87$$

$$P(87 \le Y \le 153) = .90$$

d. $P(100 \le Y \le 140) = P(-1 \le Z \le +1) = .3413 + .3413 = .6826 = 68.3\%$

e. $P(60 \le Y \le 180) = P(-3 \le Z \le +3) = .49865 + .49865 = .9973 = 99.7\%$

   $< 1\%$ of the observations lie outside the interval.

5. The mean of the population of mean values is equal to 120 (mmHg) and the standard deviation is

$$\frac{\sigma}{\sqrt{n}} = \frac{20}{\sqrt{100}} = 2$$

a. $P(\bar{Y} \ge 126) = P\left[Z \ge \dfrac{126 - 120}{2}\right] = P(Z \ge 3.0) = .5 - .49865 = .001$

b. $Z = \dfrac{\bar{Y} - \mu}{\sigma/\sqrt{n}} = \dfrac{126 - 120}{2} = 3.0$

c. 95% of all sample means fall within ± 1.96 standard deviations of μ.

$$Z = \frac{\bar{Y} - \mu}{\sigma/\sqrt{n}} \qquad -1.96 = \frac{\bar{Y} - 120}{2}$$

$$\bar{Y} = 116.08$$

$$1.96 = \frac{\bar{Y} - 120}{2}$$

$$\bar{Y} = 123.92$$

$$P(116.08 \le \bar{Y} \le 123.92) = .95$$

d. The mean value for medical students ($\bar{Y} = 126$) falls outside the range in which we would expect 95% of the mean values to fall. It appears that the population of blood pressure readings for medical students has a mean μ that is greater than the mean for all males aged 20–24.

e. The Z value above which 10% of the means will fall is $+1.28$.

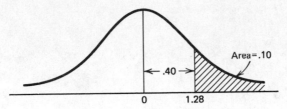

Area=.10

.40

0   1.28

$$Z = \frac{\bar{Y} - \mu}{\sigma/\sqrt{n}}$$ 10% of the mean values will lie above $\bar{Y} = 122.56$.

$$1.28 = \frac{\bar{Y} - 120}{2}$$

$$\bar{Y} = 122.56$$

f. $P(\bar{Y} \leq 115) = P\left(Z \leq \frac{115 - 120}{2}\right) = P(Z \leq -2.5) = .5 - .4938 = .0062$

6. a. $CI_\mu = \bar{Y} \pm t_{.025} \dfrac{s}{\sqrt{n}}$ $df = n - 1 = 9 - 1 = 8$

   $= 7.5 \pm (2.306) \dfrac{6}{\sqrt{9}}$ $\alpha = .05, \dfrac{\alpha}{2} = .025$

   $= [2.89, 12.11]$

b. $CI_\mu = 6.5 \pm (2.131) \dfrac{2}{\sqrt{16}}$ $df = 15$

   $= [5.43, 7.57]$ $\alpha = 0.5$

7. a. $CI_\mu = \bar{Y} \pm Z_{.475} \dfrac{\sigma}{\sqrt{n}} = 260 \pm (1.96) \dfrac{40}{\sqrt{100}} = [252.16, 267.84]$

b. Since the true mean of the normal males is not contained in the interval, it appears that true mean cholesterol levels are different for the two groups.

8. $CI = \bar{Y} \pm Z_{.495} \dfrac{\sigma}{\sqrt{n}} = 126 \pm (2.58) \dfrac{20}{\sqrt{100}} = [120.84, 131.16]$

CHAPTER 4 SOLUTIONS

1. STEP 1. Hypotheses:

   $H_0: \mu = 210$
   $H_a: \mu < 210$

STEP 2. Sample values:

$\bar{Y} = 170$ Given: $\sigma = 20$

$n = 25$

STEP 3. Test statistic:

$$Z = \frac{\bar{Y} - \mu}{\sigma/\sqrt{n}} = \frac{170 - 210}{20/\sqrt{25}} = -10.$$

STEP 4. Rejection region (lower-tail test):

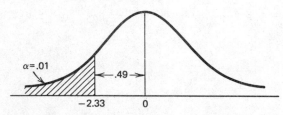

STEP 5. Conclusion: The null hypothesis is rejected at the $\alpha = .01$ level of significance, and it may be concluded that premature infants have a lower mean urine chloride level than the clinical norm.

2. $CI_\mu = \bar{Y} \pm Z_{.495}\dfrac{\sigma}{\sqrt{n}} = 170 \pm 2.58\dfrac{20}{\sqrt{25}} = [159.68, 180.32]$

3. STEP 1. Hypotheses (two-tailed test):

$H_0: \mu = 375$
$H_a: \mu \neq 375$

STEP 2. Sample values:

$\bar{Y} = 450$ $s = 62.5$

$n = 9$ $\dfrac{s}{\sqrt{n}} = \dfrac{62.5}{\sqrt{9}} = 20.833$

STEP 3. Test statistic:

$$t = \frac{\bar{Y} - \mu}{s/\sqrt{n}} = \frac{450 - 375}{20.833} = 3.6$$

STEP 4. Rejection region ($\alpha = .05$):

$df = n - 1 = 8$

STEP 5. Conclusion: The null hypothesis is rejected at the $\alpha = .05$ level of significance. It may be concluded that the mean amount of Glucose-6-dehydrogenase in the blood is different for heroin addicts than for normal adult males.

4. $CI_\mu = \bar{Y} \pm t_{.025} \dfrac{s}{\sqrt{n}} = 450 \pm 2.306(20.833) = [401.96, 498.04]$

5. STEP 1. Hypotheses:

$$H_0: \mu_1 - \mu_2 = 0$$
$$H_a: \mu_1 - \mu_2 \neq 0$$

STEP 2. Sample values:

Propranolol	Control
$n_1 = 12$	$n_2 = 11$
$\bar{Y}_1 = 120$	$\bar{Y}_2 = 70$
$s_1 = 10$	$s_2 = 8$

$$s_p = \sqrt{\frac{11(10)^2 + 10(8)^2}{21}} = 9.103$$

$$s_p \sqrt{\frac{1}{n_1} + \frac{1}{n_2}} = 9.103 \sqrt{\frac{1}{12} + \frac{1}{11}} = 3.799$$

STEP 3. Test statistic (pooled t):

$$t = \frac{\bar{Y}_1 - \bar{Y}_2}{s_p \sqrt{\dfrac{1}{n_1} + \dfrac{1}{n_2}}} = \frac{120 - 70}{3.799} = 13.2$$

STEP 4. Rejection region:

$$\alpha = .05 \qquad df = n_1 + n_2 - 2 = 21$$

$$\frac{\alpha}{2} = .025 \qquad\qquad \frac{\alpha}{2} = .025$$

$$-2.08 \qquad\qquad 2.08$$

STEP 5. Conclusion: The null hypothesis is rejected at the $\alpha = .05$ level of significance, and it may be concluded that mean cumulative weight loss is different for the propranolol and control groups.

6. $CI_{(\mu_1 - \mu_2)} = (\bar{Y}_1 - \bar{Y}_2) \pm t_{.025} s_p \sqrt{\dfrac{1}{n_1} + \dfrac{1}{n_2}}$

$$= (120 - 70) \pm (2.08)(3.799)$$

$$= [42.1, 57.9]$$

7.

	Serum Digoxin Concentration		
Subject	*4 Hours*	*8 Hours*	*Difference*
1	1.0	1.0	0
2	1.3	1.3	0
3	0.9	0.7	−0.2
4	1.0	1.0	0
5	1.0	0.9	−0.1
6	0.9	0.8	−0.1
7	1.3	1.2	−0.1
8	1.1	1.0	−0.1
9	1.0	1.0	0

STEP 1.   Hypotheses:

$$H_0 : D = 0$$
$$H_a : D \neq 0$$

STEP 2.   Sample values:

$$\bar{d} = -0.067 \qquad s_d = .071$$

$$\frac{s_d}{\sqrt{n}} = \frac{.071}{\sqrt{9}} = 0.024$$

STEP 3.   Test statistic (paired t):

$$t = \frac{\bar{d} - 0}{s_d / \sqrt{n}} = \frac{-0.067}{.024} = -2.8$$

STEP 4.   Rejection region ($\alpha = .05$):

$$df = n - 1 = 8$$

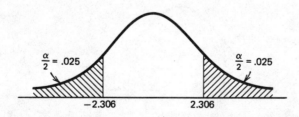

$\frac{\alpha}{2} = .025$ $\frac{\alpha}{2} = .025$

−2.306   2.306

STEP 5.   Conclusion: The null hypothesis is rejected at the $\alpha = .05$ level of significance. It may be concluded that a statistically significant difference in serum digoxin concentration does exist for the 4 hours and 8 hours periods.

8. $CI_D = \bar{d} \pm t_{.025} \dfrac{s_d}{\sqrt{n}} = -0.067 \pm 2.306(.024) = [-.01, -.12]$

9. STEP 1. Hypotheses:

$$H_0: \mu_1 - \mu_2 = 0$$
$$H_a: \mu_1 \text{ (rapid injection)} > \mu_2 \text{ (infusion)}$$

or

$$\mu_1 - \mu_2 > 0$$

STEP 2. Sample values (measurement made at end of 8 hours):

Injection	*Infusion*
$\bar{Y}_1 = .989$	$\bar{Y}_2 = .950$
$n_1 = 9$	$n_2 = 11$
$s_1 = .183$	$s_2 = .2$

$$s_p = \sqrt{\frac{8(.183)^2 + 10(.2)^2}{18}} = .193$$

$$s_p = \sqrt{\frac{1}{n_1} + \frac{1}{n_2}} = .193\sqrt{\frac{1}{9} + \frac{1}{11}} = .087$$

STEP 3. Test statistic (pooled t):

$$t = \frac{\bar{Y}_1 - \bar{Y}_2}{s_p\sqrt{\dfrac{1}{n_1} + \dfrac{1}{n_2}}} = \frac{.989 - .950}{.087} = .45$$

STEP 4. Rejection region ($\alpha = .01$) (upper-tail test):

$$df = n_1 + n_2 - 2 = 18$$

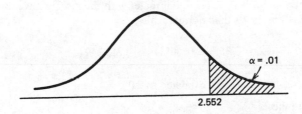

$\alpha = .01$

2.552

STEP 5. Conclusion: The null hypothesis is not rejected. There is insufficient evidence to conclude that the serum digoxin level is higher 8 hours following rapid injection than at the end of 8 hours following intravenous infusion.

10. $\text{CI}_{\mu_1 - \mu_2} = (\bar{Y}_1 - \bar{Y}_2) \pm t_{.005}s_p\sqrt{\dfrac{1}{n_1} + \dfrac{1}{n_2}}$

$$= (.989 - .950) \pm (2.878)(.087)$$
$$= [-0.21, 0.29]$$

11. STEP 1.   Hypotheses:

$H_0: \mu = 4.6$

H_a: Males with the disease have lower plasma potassium levels than normal males. ($\mu < 4.6$)

STEP 2.   Sample values:

$n = 50$

$\bar{Y} = 3.35 \qquad \dfrac{s}{\sqrt{n}} = .071$

$s = 0.50$

STEP 3.   Test statistic:

$$t = \frac{\bar{Y} - \mu}{s/\sqrt{n}} = \frac{3.35 - 4.6}{.071} = -17.6$$

STEP 4.   Rejection region (lower tail):

$t_{.05} = 1.68$

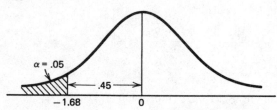

$\alpha = .05$

$\longleftarrow .45 \longrightarrow$

$-1.68 \qquad\qquad 0$

STEP 5.   Conclusion: The null hypothesis is rejected at the $\alpha = .05$ level of significance, and it is concluded that males with the disease have lower plasma potassium levels than normal males.

12. $\text{CI}_\mu = \bar{Y} \pm t_{.025} \dfrac{s}{\sqrt{n}} = 3.35 \pm 2.01(.071) = [3.21, 3.49]$

Note that $t_{.025}$, 49 df was approximated as 2.01.

13. STEP 1.   Hypotheses:

$H_0: \mu_B - \mu_C = 0$

H_a: Urinary calcium excretion is different for the two groups.
$\mu_B - \mu_C \neq 0$

STEP 2.   Sample values:

Bumetanide	*Control*
$n_1 = 9$	$n_2 = 16$
$\bar{Y}_1 = 7.5$	$\bar{Y}_2 = 6.5$
$s_1 = 6.0$	$s_2 = 2.0$

$$s_p = \sqrt{\frac{8(6.0)^2 + 15(2.0)^2}{9 + 16 - 2}} = 3.89$$

$$s_p \sqrt{\frac{1}{n_1} + \frac{1}{n_2}} = 3.89 \sqrt{\frac{1}{9} + \frac{1}{16}} = 1.62$$

STEP 3. Test statistic (pooled t):

$$t = \frac{\bar{Y}_1 - \bar{Y}_2}{s_p \sqrt{\frac{1}{n_1} + \frac{1}{n_2}}} = \frac{7.5 - 6.5}{1.62} = 0.62$$

STEP 4. Rejection region (two-tailed):

$$t_{.025} = \pm 2.069 \qquad df = n_1 + n_1 - 2 = 23$$

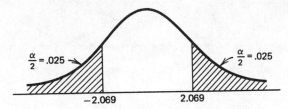

$\frac{\alpha}{2} = .025$ $\frac{\alpha}{2} = .025$

-2.069 2.069

STEP 5. Conclusion: The null hypothesis cannot be rejected. There is insufficient evidence to show that urinary calcium excretion is different for the bumetanide and control groups.

14. $\mathrm{CI}_{\mu_B - \mu_C} = (\bar{Y}_1 - \bar{Y}_2) \pm t_{.025} s_p \sqrt{\frac{1}{n_1} + \frac{1}{n_2}}$

$$= (7.5 - 6.5) \pm 2.069(1.62)$$

$$= [-2.35, 4.35]$$

15. STEP 1. Hypotheses:

$H_0: \mu = 240$

$H_a: \mu > 240$ for patients with a coronary event

STEP 2. Sample values:

$$n = 100 \qquad \frac{\sigma}{\sqrt{n}} = 4.0$$

$$\bar{Y} = 260$$

$$\sigma = 40$$

STEP 3. Test statistic:

$$Z = \frac{\bar{Y} - \mu}{\sigma/\sqrt{n}} = \frac{260 - 240}{4} = 5.0$$

STEP 4. Rejection region (upper tail):

$$Z_{.45} = 1.65$$

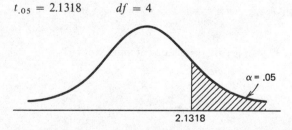

α = .05

.45

0   1.65

STEP 5. Conclusion: The null hypothesis is rejected at the $\alpha = .05$ level of significance. There is sufficient evidence to support the claim that patients with a coronary event have higher cholesterol values than normal healthy males.

16. STEP 1. Hypotheses:

$H_0: \mu = 100$

H_a: The mean number of samples analyzed per day is greater than 100.

STEP 2. Sample values:

$$n = 5 \qquad \frac{s}{\sqrt{n}} = \frac{5}{\sqrt{5}} = 2.236$$

$$\bar{Y} = 97$$

$$s = 5$$

STEP 3. Test statistic:

$$t = \frac{\bar{Y} - \mu}{s/\sqrt{n}} = \frac{97 - 100}{2.236} = -1.34$$

STEP 4. Rejection region:

$$t_{.05} = 2.1318 \qquad df = 4$$

α = .05

2.1318

STEP 5. Conclusion: There is insufficient evidence to show that the mean number of samples produced per day is greater than 100 ($\alpha = .05$). The analyzer should not be purchased.

17. STEP 1. Hypotheses:

$$H_0: \mu_1 - \mu_2 = 0$$
$$H_a: \mu_1 - \mu_2 \neq 0$$

STEP 2. Sample values:

Premature	*Full Term*
$n_1 = 5$	$n_2 = 5$
$\bar{Y}_1 = 2.0$	$\bar{Y}_2 = 4.0$

$$\sigma_1 = \sigma_2 = 1.0$$

$$\sigma \sqrt{\frac{1}{n_1} + \frac{1}{n_2}} = 1.0\sqrt{\frac{1}{5} + \frac{1}{5}} = .632$$

STEP 3. Test statistic:

$$Z = \frac{\bar{Y}_1 - \bar{Y}_2}{\sigma \sqrt{\dfrac{1}{n_1} + \dfrac{1}{n_2}}} = \frac{2.0 - 4.0}{.632} = -3.16$$

STEP 4. Rejection region ($\alpha = .01$):

$$z_{.495} = 2.58$$

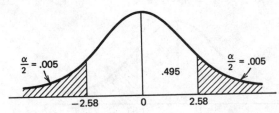

STEP 5. Conclusion: The null hypothesis is rejected at the $\alpha = .01$ level of significance. It may be concluded that bilirubin levels are different for the two groups.

18. $\text{CI}_{(\mu_1 - \mu_2)} = (\bar{Y}_1 - \bar{Y}_2) \pm z_{.495}\sigma \sqrt{\dfrac{1}{n_1} + \dfrac{1}{n_2}}$

$$= (2.0 - 4.0) \pm 2.58(.632)$$

$$= [-.37, -3.63]$$

19.   *Weight Gain in Nine Females Following*
            Contraceptive Usage

Subject	Initial	3 Months	Difference
1	120	123	+3
2	141	143	+2
3	130	140	+10
4	150	145	−5
5	135	140	+5
6	140	143	+3
7	120	118	−2
8	140	141	+1
9	130	132	+2

STEP 1.   Hypotheses:

$$H_0:D = 0$$
$$H_a:D > 0$$

STEP 2.   Sample values:

$$n = 9 \qquad \frac{s}{\sqrt{n}} = 1.399$$

$$\bar{d} = 2.111$$

$$s = 4.197$$

STEP 3.   Test statistic:

$$t = \frac{\bar{d} - 0}{s/\sqrt{n}} = \frac{2.111}{1.399} = 1.51$$

STEP 4.   Rejection region (upper-tail):

$$t_{.05} = 1.86 \qquad df = n - 1 = 8$$

$\alpha = .05$

1.86

STEP 5.   Conclusion: The null hypothesis cannot be rejected ($\alpha = .05$). There is insufficient evidence to show that women using the oral contraceptive experience a weight gain.

20. $CI_d = \bar{d} \pm t_{.025} \dfrac{s}{\sqrt{n}}$

$\qquad = 2.111 \pm 2.306(1.399)$

$\qquad = [-1.11, 5.34]$

CHAPTER 5 SOLUTIONS

1. Intermediate calculations:

a. $\sum x^2 = \sum X^2 - \dfrac{(\sum X)^2}{n} = 1002 - \dfrac{(84)^2}{9} = 218 \qquad \bar{X} = \dfrac{84}{9} = 9.3$

$\sum y^2 = \sum Y^2 - \dfrac{(\sum Y)^2}{n} = 642 - \dfrac{(72)^2}{9} = 66 \qquad \bar{Y} = \dfrac{72}{9} = 8.0$

$\sum xy = \sum XY - \dfrac{(\sum X)(\sum Y)}{n} = 780 - \dfrac{(84)(72)}{9} = 108$

b. $\hat{\beta}_1 = \dfrac{\sum xy}{\sum x^2} = \dfrac{108}{218} = .4954$

$\hat{\beta}_0 = 8.0 - (0.4954)(9.3333) = 3.3762$

Regression line: $\hat{Y} = 3.3762 + 0.4954X$

c. SSE $= \sum y^2 - \hat{\beta}_1 \sum xy$

$\qquad = 66 - (0.4954)(108)$

$\qquad = 12.4968$

$s_{y \cdot x} = \sqrt{\dfrac{\text{SSE}}{n-2}} = \sqrt{\dfrac{12.4968}{7}} = 1.3361$

$\text{CI}_{\beta_1} = \hat{\beta}_1 \pm t_{.025} \dfrac{s_{y \cdot x}}{\sqrt{\sum x^2}}$

$\qquad = .4954 \pm (2.365) \dfrac{1.3361}{\sqrt{218}}$

$\qquad = [.28, .71]$

d. STEP 1. Null hypothesis:

$\qquad H_0 : \beta_1 = 0$

STEP 2. Sample values:

$\qquad \dfrac{s_{y \cdot x}}{\sqrt{\sum x^2}} = \dfrac{1.3361}{\sqrt{218}} = .0905$

STEP 3. Test statistic:

$\qquad t = \dfrac{\hat{\beta}_1 - \beta_1}{s_{y \cdot x}/\sqrt{\sum x^2}} = \dfrac{.4954}{.0905} = 5.5$

STEP 4. Rejection region ($\alpha = .05$):

$\qquad df = n - 2 = 7$

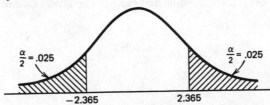

STEP 5. Conclusion: The null hypothesis of no linear relationship is rejected at the $\alpha = .05$ level of significance. It may be concluded that there is a linear relationship between dosage and sleeping time.

c. $\hat{Y} = 3.3762 + .4954X$

$\qquad = 3.3762 + .4954(12) = 9.32$ hours

2. Intermediate calculations:

a. $\sum x^2 = \sum X^2 - \dfrac{(\sum X)^2}{n} = 8675 - \dfrac{(215)^2}{6} = 970.8333$

$\sum y^2 = \sum Y^2 - \dfrac{(\sum Y)^2}{n} = 447500 - \dfrac{(1450)^2}{6} = 97083.3333$

$\sum xy = \sum XY - \dfrac{(\sum X)(\sum Y)}{n} = 51500 - \dfrac{(215)(1450)}{6} = -458.3333$

b. $r = \dfrac{\sum xy}{\sqrt{\sum x^2 \sum y^2}} = \dfrac{-458.3333}{\sqrt{(970.8333)(97083.3333)}} = -.0472$

c. STEP 1. Null hypothesis:

$H_0 : \rho = 0$

STEP 2. Sample values:

$r = -0.0472$

STEP 3. Test statistic:

$$t = r\sqrt{\dfrac{n-2}{1-r^2}} = -.095$$

STEP 4. Rejection region ($\alpha = .05$):

$t_{.025} = \pm 2.776 \qquad df = n - 2 = 4$

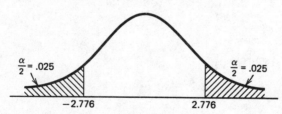

$\dfrac{\alpha}{2} = .025 \qquad\qquad \dfrac{\alpha}{2} = .025$

$-2.776 \qquad\qquad 2.776$

STEP 5. Conclusion: There is insufficient evidence to reject the null hypothesis. Therefore, it is concluded that no significant linear relationship exists between intensity rating and plasma amphetamine level.

3. Intermediate calculations:

a. $\sum x^2 = \sum X^2 - \dfrac{(\sum X)^2}{n} = 24900 - \dfrac{(450)^2}{9} = 2400$

$\sum y^2 = \sum Y^2 - \dfrac{(\sum Y)^2}{n} = 211475 - \dfrac{(1365)^2}{9} = 4450$

$\sum xy = \sum XY - \dfrac{(\sum X)(\sum Y)}{n} = 71450 - \dfrac{614250}{9} = 3200$

$\bar{Y} = \dfrac{1365}{9} = 151.6667 \qquad \bar{X} = \dfrac{450}{9} = 50$

b. $\hat{\beta}_1 = \dfrac{\sum xy}{\sum x^2} = \dfrac{3200}{2400} = 1.333333$

$\hat{\beta}_0 = 151.6667 - (1.3333)(50) = 85.00$

Regression line: $\hat{Y} = 85.00 + 1.3333X$

c. $\text{SSE} = \sum y^2 - \hat{\beta}_1 \sum xy$

$\quad\quad = 4450 - (1.3333)(3200) = 183.3344$

$s_{y\cdot x} = \sqrt{\dfrac{\text{SSE}}{n-2}} = 5.1178$

$\text{CI}_{\beta_1} = \hat{\beta}_1 \pm t_{.025} \dfrac{s_{y\cdot x}}{\sqrt{\sum x^2}}$

$\quad\quad = 1.3333 \pm (2.365)\dfrac{5.1178}{\sqrt{2400}} = [1.1, 1.6]$

d. STEP 1. Null hypothesis:

$\quad\quad H_0: \beta_1 = 0$

$\quad\quad H_a: \beta_1 \neq 0$

STEP 2. Sample values:

$\quad\quad \dfrac{s_{y\cdot x}}{\sqrt{\sum x^2}} = .1045$

STEP 3. Test statistic:

$\quad\quad t = \dfrac{\hat{\beta}_1 - \beta_1}{s_{y\cdot x}/\sqrt{\sum x^2}} = \dfrac{1.3333}{.1045} = 12.8$

STEP 4. Rejection region ($\alpha = .01$):

$\quad\quad t_{.005} = \pm 3.499 \quad\quad df = n - 2 = 7$

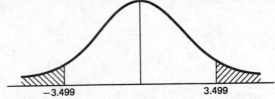

$\quad\quad -3.499 \quad\quad\quad\quad\quad\quad 3.499$

STEP 5. Conclusion: The null hypothesis of no linear relationship between variables is rejected at the $\alpha = .01$ level of significance. It is concluded that there is a significant linear relationship between stress (shock) and blood pressure in monkeys.

e. $\hat{Y} = 85 + 1.3333(60) = 164.998$ mmHg

4. Intermediate calculations:

a. $\sum x^2 = 85.1323 - \dfrac{(31.41)^2}{12} = 2.916625$

$\sum y^2 = 2816050 - \dfrac{(5720)^2}{12} = 89516.66667$

$\sum xy = 15357.55 - \dfrac{(31.41)(5720)}{12} = 385.45$

$\bar{Y} = \dfrac{5720}{12} = 476.66667 \qquad \bar{X} = \dfrac{31.41}{12} = 2.6175$

b. $r = \dfrac{\sum xy}{\sqrt{\sum x^2 \sum y^2}} = \dfrac{385.45}{\sqrt{(2.916625)(89516.66667)}} = 0.7544$

c. From Appendix Table D, the critical r value for $df = 10$ and $\alpha = .05$ is .5760. There is therefore a significant correlation between GPR and National Board scores ($\alpha = .05$).

d. $\hat{\beta}_1 = \dfrac{\sum xy}{\sum x^2} = \dfrac{385.45}{2.916625} = 132.1562$

$\hat{\beta}_0 = \bar{Y} - \hat{\beta}_1\bar{X} = 476.66667 - (132.1562)(2.6175) = 130.7479$

$\hat{Y} = 130.7479 + 132.1562X$

e. $\hat{Y} = 527.2 \quad \text{for} \quad X = 3$

5. Intermediate calculations:

a. $\sum x^2 = 60.5 - \dfrac{(21)^2}{10} = 16.4$

$\sum y^2 = .8526 - \dfrac{(2.74)^2}{10} = .10184$

$\sum xy = 4.535 - \dfrac{(21)(2.74)}{10} = -1.219$

$\bar{X} = \dfrac{21}{10} = 2.1 \qquad \bar{Y} = \dfrac{2.74}{10} = .274$

b. $\hat{\beta}_1 = \dfrac{\sum xy}{\sum x^2} = \dfrac{-1.219}{16.4} = -.07433$

$\hat{\beta}_0 = \bar{Y} - \hat{\beta}_1\bar{X} = .274 - (-.07433)(2.1) = .4301$

Regression line: $\hat{Y} = .4301 - .07433X$

c. $\text{SSE} = \sum y^2 - \hat{\beta}_1 \sum xy$

$= 0.10184 - (-0.07433)(-1.219)$

$= 0.01123$

$$s_{y \cdot x} = \sqrt{\frac{SSE}{8}} = 0.03747 \qquad \frac{s_{y \cdot x}}{\sqrt{\sum x^2}} = 0.00925$$

$$CI_{\beta_1} = \hat{\beta}_1 \pm t_{.005} \frac{s_{y \cdot x}}{\sqrt{\sum x^2}}$$

$$= -0.07433 \pm 3.355(.00925)$$

$$= [-.11, -.04]$$

d. STEP 1. Null hypothesis:

$$H_0: \beta_1 = 0$$
$$H_a: \beta_1 \neq 0$$

STEP 2. Sample values:

$$\hat{\beta}_1 = .07433$$

$$\frac{s_{y \cdot x}}{\sqrt{\sum x^2}} = .00925$$

STEP 3. Test statistic:

$$t = \frac{\hat{\beta}_1 - \beta_1}{s_{y \cdot x}/\sqrt{\sum x^2}} = \frac{-.07433}{.00925} = -8.04$$

STEP 4. Rejection region ($\alpha = .01$):

$$t_{.005} = \pm 3.355 \qquad df = n - 2 = 8$$

STEP 5. Conclusion: The null hypothesis of no linear relationship is rejected at the $\alpha = .01$ level of significance.

e. $\hat{Y} = .28 \ \mu g/ml$ after 2 hours

CHAPTER 6 SOLUTIONS

1.

	Drug			
	A	B	C	D
	10	12	9	17
	8	14	13	14
	7	11	10	13
	11	15	12	16
$\sum Y$:	36	52	44	60
$\sum Y^2$:	334	686	494	910
$\bar{Y}$:	9.0	13.0	11.0	15
s:	1.826	1.826	1.826	1.826
S.E.:	0.91	0.91	0.91	0.91

a. Completely randomized design (CRD)
b. See table above.

c. $SSE_A = 10^2 + 8^2 + 7^2 + 11^2 - \dfrac{(36)^2}{4} = 10$

$SSE_B = 12^2 + 14^2 + 11^2 + 15^2 - \dfrac{(52)^2}{4} = 10$

$SSE_C = 9^2 + 13^2 + 10^2 + 12^2 - \dfrac{(44)^2}{4} = 10$

$SSE_D = 17^2 + 14^2 + 13^2 + 16^2 - \dfrac{(60)^2}{4} = 10$

$SSE = SSE_A + SSE_B + SSE_C + SSE_D = 40$

$SST = \dfrac{(36)^2}{4} + \dfrac{(52)^2}{4} + \dfrac{(44)^2}{4} + \dfrac{(60)^2}{4} - \dfrac{(192)^2}{16} = 80$

$SS_{Total} = SST + SSE = 80 + 40 = 120$

ANOVA

Source	df	SS	MS	F
Among treatments	3	80	26.7	8.1
Within treatments	12	40	3.3	
Total	15	120		

Tabulated $F_{3,12} = 3.49$ ($\alpha = .05$). Since the calculated F is larger than the tabulated F, it may be concluded that the drugs have different effects on changes in blood pressure (at the $\alpha = .05$ level of significance).

d. $SS_{Total} = 10^2 + 8^2 + \cdots + 16^2 - \dfrac{(192)^2}{16} = 120$

     $SSE = 120 - 80 = 40$

Same as before.

e. From the significant F in the ANOVA, the size of the standard errors of the means, and the value of the means, it would appear that there is a steady, significant increase in effect in the drug order A, C, B, D.

2. a. Randomized complete block, each enzyme acts as a block.
  b. Intermediate calculations:

		Block			
Lab	1	2	3	4	Total
A	4.2	6.0	3.9	8.3	22.4
B	3.9	7.3	4.0	7.2	22.4
C	5.2	6.5	3.2	6.9	21.8
Total	13.3	19.8	11.1	22.4	66.6

Correction factor $= \dfrac{(\sum Y_i)^2}{N} = \dfrac{(66.6)^2}{12} = 369.63$

$SSB = \dfrac{(13.3)^2 + (19.8)^2 + (11.1)^2 + (22.4)^2}{3} - 369.63$

$= 397.97 - 369.63$

$= 28.32$

$SST = \dfrac{(22.4)^2 + (22.4)^2 + (21.8)^2}{4} - 369.63$

$= 369.69 - 369.63 = 0.06$

$SS_{Total} = 4.2^2 + 3.9^2 + 5.2^2 + 6.0^2 + \cdots + 6.9^2 - 369.63$

$= 401.22 - 369.63 = 31.59$

$SSE = 31.59 - 28.34 - 0.06 = 3.19$

ANOVA

Source of Variation	Degrees of Freedom	Sum of Squares	Mean Square	F
Blocks	3	28.34		
Treatments	2	0.06	0.03	0.06
Error	6	3.19	0.53	
Total	11			

3. a. Completely randomized design.

b. *Treatment Group*

1	2	3
4	3	12
7	5	8
6	2	9
3		11
2		
(22)	(10)	(40)

$CF = \dfrac{(72)^2}{12} = 432$

$SST = \dfrac{(22)^2}{5} + \dfrac{(10)^2}{3} + \dfrac{(40)^2}{4} - 432$

$= 530.13 - 432. = 98.1$

$SS_{Total} = 4^2 + 7^2 + 6^2 + \cdots + 11^2 - 432 = 130$

$SSE - 130 - 98.1 = 31.9$

ANOVA

Source of Variation	Degrees of Freedom	Sum of Squares	Mean Squares	F
Treatment	2	98.1	49.05	13.8
Error	9	31.9	3.54	
Total	11	130.0		

Tabulated $F_{2,9} = 8.02$; thus, we may conclude that the treatments are different at the $\alpha = .01$ level of significance.

4.

Treatments

Blocks	1	2	3	4	Total
1	4.0	3.1	4.4	5.9	(17.4)
2	6.6	6.4	3.3	1.9	(18.2)
3	4.9	7.1	4.0	4.0	(20)
4	7.3	6.7	6.8	3.1	(23.9)
(Total)	(22.8)	(23.3)	(18.5)	(14.9)	(79.5)

$$SSB = \frac{(17.4)^2 + (18.2)^2 + (20)^2 + (23.9)^2}{4} - \frac{(79.5)^2}{16}$$

$$= 401.3 - 395.02 = 6.3$$

$$SST = \frac{(22.8)^2 + (23.3)^2 + (18.5)^2 + (14.9)^2}{4} - \frac{(79.5)^2}{16}$$

$$= 406.7 - 395.02 = 11.7$$

$$SS_{Total} = 4^2 + 6.6^2 + 7.3^2 + \cdots + 3.1^2 - \frac{(79.5)^2}{16} = 44.25$$

$$SSE = 44.25 - 11.7 - 6.3 = 26.2$$

ANOVA

Source of Variation	Degrees of Freedom	Sum of Squares	Mean Squares	F
Blocks	3	6.30		
Treatments	3	11.75	3.92	1.34
Error	9	26.20	2.91	
Total	15	44.25		

Tabulated $F_{3,9} = 6.99$; we do not have evidence that the treatments are different at the $\alpha = .01$ level of significance.

5. a.

	Motrin, 400 mg	Codeine, 60 mg	Codeine, 30 mg	Placebo	
	82	80	77	65	
	89	70	69	75	
	77	72	67	67	
	72	90	65	55	
	92	68	57	63	Total
Total (T_i)	412	380	335	325	1452
n_i	5	5	5	5	20
$\bar{Y}_i$	82.4	76	67	65	72.6

Calculation of sum of squares (completely random design):

$$CF = \frac{(\sum Y)^2}{N} = \frac{(1452)^2}{20} = 105415.2$$

$$SS_{Total} = \sum_{all} Y_i^2 - CF = 107416 - 105415.2 = 2000.8$$

$$SS_{Among\ treatments} = SST = \sum_{i=1}^{4} \frac{(T_i)^2}{n_i} - CF$$

$$= \frac{(412)^2 + (380)^2 + (335)^2 + (325)^2}{5} - 105415.2$$

$$= 106398.8 - 105415.2 = 983.6$$

$$SSE = SS_{Total} - SST = 2000.8 - 983.6 = 1017.2$$

ANOVA

Source	df	SS	MS	F
Treatments	3	983.6	327.867	5.16
Error	16	1017.2	63.575	
Total	19	2000.8		

Tabulated $F_{3,16,.05} = 3.24$. Since the calculated $F = 5.16$ is greater than the tabulated value (3.24), we may reject the null hypothesis of equality of treatment effects. We may conclude at the $\alpha = .05$ level of significance that at least one pair of population treatment means are not equal. The data do provide evidence to indicate a difference in perceived pain relief among the four treatments.

b. There are six possible pairwise comparisons among the four treatments.

Hypothesis	t Statistic
$\mu_1 = \mu_2$	$\dfrac{82.4 - 76}{\sqrt{(63.575)\left(\dfrac{1}{5} + \dfrac{1}{5}\right)}} = 1.27$
$\mu_1 = \mu_3$	$\dfrac{82.4 - 67}{5.04} - 3.05$

Hypothesis	t Statistic
$\mu_1 = \mu_4$	$\dfrac{82.4 - 65}{5.04} = 3.45$
$\mu_2 = \mu_3$	$\dfrac{76 - 67}{5.04} = 1.78$
$\mu_2 = \mu_4$	$\dfrac{76 - 65}{5.04} = 2.18$
$\mu_3 = \mu_4$	$\dfrac{67 - 65}{5.04} = .40$

To preserve the $\alpha = .05$ per experiment error rate, we compare each of the t statistics above with $t_{(.05/6)/2} = t_{.004}$ for the two-tailed alternative $\mu_i \neq \mu_j$.

$$t_{.004.16} = z_{.496} + \frac{z_{.496}^3 + z_{.496}}{4(14)} = 2.65 + \frac{(2.65)^3 + 2.65}{4(14)} = 3.03$$

Thus, we conclude that Motrin (400 mg) is simultaneously different from codeine (30 mg) and from placebo.

6. a. Randomized complete block design.

Treatments

Blocks Patient	Diet Alone	Chlorpropamide (100 mg/day)	Chlorpropamide (250 mg/day)	Block Totals (B_i)
1	8	5	5	18
2	7	6	5	18
3	9	8	7	24
4	7	5	5	17
5	8	6	7	21
Treatment totals (T_i)	39	30	29	98
$\overline{Y}_i$	7.8	6.0	5.8	—

Calculation of sums of squares:

$$CF = \frac{(\sum Y)^2}{N} = \frac{(98)^2}{15} = 640.267$$

$$SS_{Total} = \sum_{all} Y^2 - CF = 666 - 640.267 = 25.733$$

$$SS_{Treatments} = SST = \frac{(39)^2 + (30)^2 + (29)^2}{5} - 640.267$$

$$= 652.4 - 640.267 = 12.133$$

$$SS_{Blocks} = SSB = \frac{(18)^2 + (18)^2 + (24)^2 + (17)^2 + (21)^2}{3} - 640.267$$

$$= 651.333 - 640.267 = 11.067$$

$$SSE = 25.7333 - 12.133 - 11.067 = 2.533$$

ANOVA

Source	df	SS	MS	F
Treatments	2	12.133	6.067	19.16
Blocks	4	11.067	2.767	
Error	8	2.533	.317	
Total	14	25.733		

The tabulated value is $F_{2.8,.01} = 8.65$. Since the calculated $F = 19.16$ is greater than the tabulated value, we reject the null hypothesis of equality of treatment effects at the $\alpha = .01$ level of significance. We may conclude that at least one pair of population means is different. The data do provide sufficient evidence to indicate a difference in Hb A_{1c} (percentage) among the three different treatments.

b. *Hypothesis*   *t Statistic*

$\mu_1 = \mu_2$ $\dfrac{7.8 - 6.0}{\sqrt{.317\left(\dfrac{1}{5} + \dfrac{1}{5}\right)}} = 5.05$

$\mu_1 = \mu_3$ $\dfrac{7.8 - 5.8}{.356} = 5.62$

$\mu_2 = \mu_3$ $\dfrac{6.0 - 5.8}{.356} = .562$

$t_{(.01/3)/2} = t_{.002}$

$$t_{.002,8} = z_{.498} + \frac{z_{.498}^3 + z_{.498}}{4(6)} = 2.88 + \frac{(2.88)^3 + 2.88}{24} = 4.0$$

We may conclude that diet alone is different from chlorpropamide (100 mg/day) and from chlorpropamide (250 mg/day) at the overall per experiment error rate of $\alpha = .01$.

7. Completely random design.

	Drug A	Drug B	Drug C	
	25	20	25	
	15	16	15	
	20	18	20	
	14	25	20	
T_i	74	79	80	233
n_i	4	4	4	12
$\bar{Y}_i$	18.5	19.75	20	19.417

Calculation of sums of squares:

$$CF = \frac{(\sum Y)^2}{N} = \frac{(233)^2}{12} - 4524.083$$

$$SS_{Total} = \sum_{all} Y^2 - CF = 4701 - 4524.083 = 176.917$$

$$SST = \frac{(74)^2 + (79)^2 + (80)^2}{4} - 4524.083 = 4529.25 - 4524.083 = 5.167$$

$$SSE = 176.917 - 5.167 = 171.75$$

ANOVA

Source	df	SS	MS	F
Treatments	2	5.167	2.583	.135
Error	9	171.75	19.083	
Total	11	176.917		

The tabulated F value is $F_{2,9,.05} = 4.26$. Since the calculated F (.165) is less than the tabulated F, we cannot reject the null hypothesis of equality of treatment effects at the $\alpha = .05$ level. The data do not provide sufficient evidence that there is a difference in scores for the 3 drugs.

8. a. Randomized complete block design.

Block Initial Level	Drug A	Treatment Drug B	Drug C	Block Totals (B_i)
1	35	30	25	90
2	40	25	20	85
3	25	25	20	70
4	30	25	25	80
Treatment totals (T_i)	130	105	90	325
$\bar{Y}_i$	32.5	26.25	22.5	

Calculation of sums of squares:

$$CF = \frac{(\sum Y)^2}{N} = \frac{(325)^2}{12} = 8802.083$$

$$SS_{Total} = 35^2 + 30^2 + \cdots + 25^2 + 25^2 - 8802.083 = 372.917$$

$$SST = \frac{(130)^2 + (105)^2 + (90)^2}{4} - 8802.083 = 9006.25 - 8802.083 = 204.167$$

$$SSB = \frac{(90)^2 + (85)^2 + (70)^2 + (80)^2}{3} - 8802.083$$

$$= 8875 - 8802.083 = 72.917$$
$$SSE = 372.917 - 204.167 - 72.917 = 95.833$$

ANOVA

Source	df	SS	MS	F
Treatments	2	204.167	102.083	6.39
Blocks	3	72.917	24.306	
Error	6	95.833	15.972	
Total	11	372.917	—	

The tabulated value for F is $F_{2,6,.05} = 5.14$. Since the calculated F exceeds the tabulated F, we may reject the null hypothesis of equality of treatment effects at the $\alpha = .05$ level of significance. There is sufficient evidence to conclude that there exists a difference in anxiety depression scores for the three drugs.

b. *Hypothesis*   *t Statistic*

$\mu_1 = \mu_2$ $\dfrac{32.5 - 26.25}{\sqrt{15.972\left(\dfrac{1}{4} + \dfrac{1}{4}\right)}} = 2.21$

$\mu_1 = \mu_3$ $\dfrac{32.5 - 22.5}{2.83} = 3.54$

$\mu_2 = \mu_3$ $\dfrac{26.25 - 22.5}{2.83} = 1.33$

$t_{(.05/3)2} = t_{.008}$

$t_{.008,6} = z_{.492} + \dfrac{z_{.492}^3 + z_{.492}}{4(4)} = 2.41 + \dfrac{(2.41)^3 + 2.41}{16} = 3.44$

We may conclude that drug A is different from drug C at the overall $\alpha = .05$ per experiment error rate.

CHAPTER 7 SOLUTIONS

1. One would expect to find 10 unsuccessful operations out of 100 if the success rate is 90%.

STEP 1. Null hypothesis:

$H_0 : P = .9$

That is, the true proportion of successes is equal to .9.

STEP 2. Sample values:

	Observed	Expected
Successful	85	90
Not successful	15	10
	100	100

STEP 3. Test statistic:

$$\chi^2 = \sum \frac{(|O - E| - .5)^2}{E}$$

$$= \frac{(|85 - 90| - 0.5)^2}{90} + \frac{(|15 - 10| - 0.5)^2}{10} = 2.25$$

STEP 4. Rejection region: From Table A the tabulated χ^2 for $\alpha = .05$ and 1 df is 3.84.

STEP 5. Since the calculated χ^2 is less than the tabulated χ^2, the null hypothesis cannot be rejected (at the $\alpha = .05$ level of significance). It cannot be shown that the true proportion of successes is different from .90, the national experience.

A 95% confidence interval on the true population proportion of successful operations is given by

$$\hat{p} \pm 1.96 \sqrt{\frac{\hat{p}(1 - \hat{p})}{n}}$$

$$0.85 \pm 1.96 \sqrt{\frac{(.15)(.85)}{100}}$$

$$[0.78, 0.92]$$

Interpretation—We are 95% confident that the above interval covers the true population proportion of successful operations.

2. STEP 1. Null hypothesis: There is no association between instruction and number of new cavities at the end of a 6-month period.

$$H_0 : P_{11} = P_{21}; P_{12} = P_{22}; P_{13} = P_{23}$$

(i.e., the proportion falling into the different cavity classifications is constant for the instruction and no instruction groups.)

STEP 2. Sample values:

	Number of New Cavities			
	0–1	2–3	4–5	(Total)
Instruction	30 (25)	15 (15)	5 (10)	50
No instruction	20 (25)	15 (15)	15 (10)	50
(Total)	50	30	20	100

STEP 3. Test statistic:

$$\chi^2 = \sum \frac{(O - E)^2}{E}$$

$$= \frac{(30 - 25)^2}{25} + \frac{(15 - 15)^2}{15} + \frac{(5 - 10)^2}{10} + \frac{(20 - 25)^2}{25}$$

$$+ \frac{(15 - 15)^2}{15} + \frac{(15 - 10)^2}{10} = 7.0$$

STEP 4. Rejection region: For $\alpha = .05$ and $df = 2$, the critical χ^2 from Table A is 5.99.

STEP 5. Conclusion: Since the calculated χ^2 is greater than the table value, the null hypothesis is rejected at the $\alpha = .05$ level of significance. It may be concluded that there is an association between instruction in dental hygiene and number of new cavities at the end of a 6-month period.

3. STEP 1. Null hypothesis: The proportion of females with hypertension is the same for the group using oral contraceptives and the group using other contraceptive devices.

$$H_0 : P_1 = P_2$$

STEP 2. Sample values:

	Hypertension	No Hypertension	(Total)
Oral contraceptive	8 (9.2)	32 (30.8)	40
Others	15 (13.8)	45 (46.2)	60
	23	77	100

STEP 3. Test statistic:

$$\chi^2 = \sum \frac{(|O - E| - .5)^2}{E}$$

$$= \frac{(|8 - 9.2| - .5)^2}{9.2} + \frac{(|32 - 30.8| - .5)^2}{30.8} + \frac{(|15 - 13.8| - .5)^2}{13.8}$$

$$+ \frac{(|45 - 46.2| - .5)^2}{46.2}$$

$$= .12$$

STEP 4. Rejection region: For $\alpha = .01$ and $1\ df$, the χ^2 from Table A is 6.63.

STEP 5. Conclusion: The null hypothesis is not rejected. There is insufficient evidence to show that the proportion of hypertensive females is different for oral contraceptive users and for those using other methods.

The 99% confidence interval on the true difference in proportion hypertensive for the two groups is

$$\hat{p}_1 - \hat{p}_2 \pm 2.58 \sqrt{\frac{\hat{p}_1(1 - \hat{p}_1)}{n_1} + \frac{\hat{p}_2(1 - \hat{p}_2)}{n_2}}$$

$$(0.20 - 0.25) \pm 2.58 \sqrt{\frac{(.20)(.80)}{40} + \frac{(.25)(.75)}{60}}$$

$$[-0.27, 0.17]$$

4. STEP 1. Null hypothesis: The proportion of mistakes is constant for the three treatments.

$$H_0: P_{11} = P_{21} = P_{31}$$
$$P_{12} = P_{22} = P_{32}$$
$$P_{13} = P_{23} = P_{33}$$

STEP 2. Sample values:

	Total Number of Mistakes			
	0–5	6–10	11–15	(Total)
Stimulant	10 (13.33)	20 (16.67)	20 (20)	50
Depressant	5 (13.33)	15 (16.67)	30 (20)	50
Placebo	25 (13.33)	15 (16.67)	10 (20)	50
(Total)	40	50	60	150

STEP 3. Test statistic:

$$\chi^2 = \sum \frac{(O - E)^2}{E}$$

$$= \frac{(10 - 13.33)^2}{13.33} + \frac{(20 - 16.67)^2}{16.67} + \frac{(20 - 20)^2}{20}$$

$$+ \frac{(5 - 13.33)^2}{13.33} + \frac{(15 - 16.67)^2}{16.67} + \frac{(30 - 20)^2}{20}$$

$$+ \frac{(25 - 13.33)^2}{13.33} + \frac{(15 - 16.67)^2}{16.67} + \frac{(10 - 20)^2}{20} = 27.75$$

STEP 4. Rejection region: For $\alpha = .05$ and $df = 4$, the critical χ^2 value from Table A is 9.49.

STEP 5. Conclusion: The null hypothesis is rejected at the $\alpha = .05$ level of significance. It may be concluded that the proportion of mistakes is not constant; there is an association between treatment received and total number of mistakes made.

5. STEP 1. Null hypothesis: There is no association between course difficulty rating and class rank.

STEP 2. Sample values:

Class Rank	Not Difficult	Intermediate Difficulty	Very Difficult	(Total)
Upper	10 (17.085)	40 (30.15)	17 (19.765)	67
Middle	20 (17.085)	35 (30.15)	12 (19.765)	67
Lower	21 (16.83)	15 (29.70)	30 (19.47)	66
(Total)	51	90	59	200

STEP 3. Test statistic:

$$\chi^2 = \sum \frac{(O - E)^2}{E}$$

$$= \frac{(10 - 17.085)^2}{17.085} + \frac{(40 - 30.15)^2}{30.15} + \frac{(17 - 19.765)^2}{19.765}$$

$$+ \frac{(20 - 17.085)^2}{17.085} + \frac{(35 - 30.15)^2}{30.15} + \frac{(12 - 19.756)^2}{19.765}$$

$$+ \frac{(21 - 16.83)^2}{16.83} + \frac{(15 - 29.70)^2}{29.70} + \frac{(30 - 19.47)^2}{19.47} = 24.87$$

STEP 4. Rejection region: For $\alpha = .05$ and $df = 4$, the critical χ^2 value from Table A is 9.49.

STEP 5. Conclusion: The null hypothesis is rejected at the $\alpha = .05$ level of significance. It may be concluded that there is an association between perceived difficulty and class rank.

6. STEP 1. Null hypothesis: The proportion of patients with satisfactory responses is the same for the two treatment groups.

$$H_0: P_1 = P_2$$

STEP 2. Sample values:

	Satisfactory	Not Satisfactory	(Total)
Antidepressant	12 (10)	18 (20)	30
Antidepressant and alcohol	8 (10)	22 (20)	30
(Total)	20	40	60

STEP 3. Test statistic:

$$\chi^2 = \sum \frac{(|O - E| - .5)^2}{E}$$

$$= \frac{(|12 - 10| - .5)^2}{10} + \frac{(|18 - 20| - .5)^2}{20} + \frac{(|8 - 10| - .5)^2}{10}$$

$$+ \frac{(|22 - 20| - .5)^2}{20} = 0.675$$

STEP 4. Rejection region: From Table A the tabulated χ^2 for $\alpha = .05$ and 1 df is 3.84.

STEP 5. Conclusion: The calculated χ^2 is less than the critical χ^2; therefore, the null hypothesis of equal proportion of satisfactory responses for the two treatment groups is not rejected at the $\alpha = .05$ level of significance.

7. STEP 1. Null hypothesis: There is no difference in proportions of births for institutionalized schizophrenics according to seasons.

$$H_0: P_1 = P_2 = P_3 = P_4 = .25$$

STEP 2. Sample values:

Season of Birth	Observed	Expected
Fall	20	25
Winter	35	25
Spring	20	25
Summer	25	25
	100	100

STEP 3. Test statistic:

$$\chi^2 = \sum \frac{(O - E)^2}{E}$$

$$= \frac{(20 - 25)^2}{25} + \frac{(35 - 25)^2}{25} + \frac{(20 - 25)^2}{25} + \frac{(25 - 25)^2}{25} = 6.0$$

STEP 4. Rejection region: From Table A the χ^2 for $\alpha = .05$ and for 3 df is 7.815.

STEP 5. Conclusion: Since the calculated χ^2 is less than the critical χ^2 from Table A, the null hypothesis of equal proportion of births cannot be rejected. It cannot be shown that there is a difference in number of birthdays according to different seasons of the year for institutionalized schizophrenics.

8. STEP 1. Null hypothesis: There is no association between reasons given for not sleeping well and type of treatment received.

SEEP 2. Sample values:

Reason for Not Sleeping Well	Hypnotic A	Hypnotic B	Hypnotic C	Placebo	(Total)
Restlessness	2 (3.4)	5 (4.25)	4 (3.4)	6 (5.95)	17
Woke too early	5 (5.4)	7 (6.75)	5 (5.4)	10 (9.45)	27
Trouble getting to sleep	8 (6.4)	4 (8.0)	6 (6.4)	14 (11.2)	32
Bad dreams	3 (2.4)	4 (3.0)	3 (2.4)	2 (4.2)	12
Other	2 (2.4)	5 (3.0)	2 (2.4)	3 (4.2)	12
Total	20	25	20	35	100

STEP 3. Test statistic:

$$\chi^2 = \sum \frac{(O - E)^2}{E} = 7.64$$

STEP 4. Rejection region: For $\alpha = .01$ and 12 df, the critical χ^2 value from Table A is 26.2.

STEP 5. Conclusion: The null hypothesis of no association between reason for sleeplessness and treatment cannot be rejected. There is insufficient evidence to show that perceived reason for not sleeping well is related to type of medication.

CHAPTER 9 SOLUTIONS

1. Arranging the data as follows:

Use Oral Contraceptives	Hypertensive	
	Yes	No
Yes	8	32
No	15	45

We calculate the odds ratio

$$o = \frac{8 \times 45}{15 \times 32} = .75$$

for those who use oral contraceptives. Alternatively, for those who use other than oral contraceptives, the odds ratio is

$$o = \frac{15 \times 32}{8 \times 45} = 1.33$$

Since this is a cross-sectional study, the interpretation, with respect to the population sampled, is that for every three women found to be hypertensive and using oral contraceptives, four will be found to be hypertensive and using other contraceptive methods. The chi-square is not significant (see solution to the original problem), so the association is judged not to be significantly strong.

2. Rearranging the data table we get the following:

	No. of Mistakes	
	0–10	11–15
Stimulant	30	20
Placebo	40	10

Since the focus on the study is whether mistakes *increase* with the use of stimulants, we compute the odds ratio as

$$o = \frac{20 \times 40}{30 \times 10} = 2.67$$

Thus, the use of the stimulant makes it 2.67 times more likely that a subject would score in the high category of mistakes. The corrected chi-square is 3.86, so the association is statistically significant.

3. Arranging the data as in Problem 1, Chapter 7, we may compute

$$o = \frac{85 \times 10}{15 \times 90} = \frac{85}{15} \times \frac{10}{90} = 0.63$$

Thus we see that a *rate* (e.g., 10 per 100) can be used to measure the desired association, in this case, the association of $a(n)$ (un)successful outcome with your hospital. The computed chi-square is 2.25, so the association is not significant.

INDEX